Data Analysis with RStudio

Franz Kronthaler · Silke Zöllner

Data Analysis with RStudio

An Easygoing Introduction

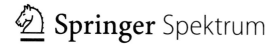 Springer Spektrum

Franz Kronthaler,
University of Applied Sciences Grisons
Chur, Switzerland

Silke Zöllner
Institue of Business and Regional
Economics IBR
Lucerne University of Applied
Sciences and Arts
Lucerne, Switzerland

ISBN 978-3-662-62517-0   ISBN 978-3-662-62518-7   (eBook)
https://doi.org/10.1007/978-3-662-62518-7

Planung: Iris Ruhmann

This Springer Spektrum imprint is published by the registered company Springer-Verlag GmbH, DE part of
Springer Nature.
The registered company address is: Heidelberger Platz 3, 14197 Berlin, Germany

Comment

R is a powerful freely available open-source tool for analyzing data and creating graphs ready for publication. In just a few years, R has become the leading statistical software in science and is now becoming even more widespread in practice and in business. R can be used to analyze data and to generate knowledge for companies and institutions that they can include in their business decisions.

The objective of the text is to introduce R—specifically RStudio—to students from different fields of study and to practitioners and enable them to use R in their everyday work. The script is not a substitute for statistical textbooks. The focus lies on the use of RStudio for data analysis, but at the same time, also some statistical knowledge is conveyed. If someone feels the need to deepen the statistical knowledge, he or she should read a textbook of statistics. At the end of the script, various textbooks are briefly described.

The main purpose however is to hand over the joy of analyzing data with RStudio!

We would like to thank Irenaeus Wolff for his critical review of the script.

Contents

List of Figures

List of Tables

R and RStudio 1

1.1 A Note on How to Use the Script

This script is designed to make it easy to get started with R and RStudio. To benefit fully from the script you should run the applications and commands shown below while working through the script itself. The required data can be found on the website www. statistik-kronthaler.ch.

The data should be saved on your computer in a working directory of your choice. This working directory is used to load the data into RStudio and to save the results. How to load data into RStudio and how to save working results is shown in detail in Chapter 2.

For even better learning results while working through the script, you can try to modify commands and test what happens, and you can apply the commands to other data sets.

To make the use of the commands as easy as possible and to design the script as clearly as possible, the commands are inserted with the usual addition of the R Console ">" at the beginning of the command. The results of the data analysis in the script are marked with the double hashmark "##".

Let's start now with some information about R and RStudio.

1.2 About R and RStudio

The development of R began in the early 1990s. R was initiated by George Ross Ihaka and Robert Gentleman (see Ihaka and Gentleman 1996). With R, they tried to use the advantages of the two programming languages S and Scheme. Since then, the project has been so successful that in just a few years R has become the standard for data analysis at universities, many companies and public authorities. As an open source solution, R does not require a license and runs on any operating system.

© The Author(s), under exclusive license to Springer-Verlag GmbH, DE, part of Springer Nature 2021
F. Kronthaler and S. Zöllner, *Data Analysis with RStudio*,
https://doi.org/10.1007/978-3-662-62518-7_1

R can be understood as a platform with which the most diverse applications of data analysis are possible. These applications are usually organized in so-called packages. For almost all needs, special analyses, exotic colors and other special requests, R can be extended with a suitable package. R can also be used to produce graphs ready for publication that can be exported directly in various file formats. R and its analysis packages are continuously developed by an R development team and the large community of users. With R, the users have a statistical software at their disposal which includes the statistical possibilities of professional and paid statistical programs but with a much higher dynamic. However, R, which is operated via the so-called console, which many people do not find very user-friendly, and additional tools have been developed to simplify the use of R. In particular, RStudio has become established.

RStudio is a standalone application based on R, where all functions available in R can be used. However, in contrast to the pure R, RStudio offers an appealing and modern interface. This makes it possible to analyze data comfortably and to prepare it graphically. RStudio offers a window for creating scripts, supports command entry and includes visualization tools. RStudio's interface is divided into four sections, providing a simultaneous overview of the data, commands, results, and graphics produced. The philosophy of RStudio, which is developed and provided by RStudio, Inc., is to empower users to use R productively.

Data is the raw material of the twenty-first century. RStudio is the tool to exploit this raw material.

It is worth to learn RStudio!

1.3   How to Install R and RStudio

To install RStudio, install R first. We can find R on the website https://www.r-project.org/ (Fig. 1.1).

We follow the link "download R" and come to a page called Cran Mirrors. A mirror is a place where R and its packages are available. If we look at the page we see that R is usually provided by universities of the respective countries, in Germany this is e.g. the University of Münster, in Switzerland the ETH Zurich or in Austria the Vienna University of Economics. We usually select the country in which we are located and go to the download area. Here you can download R for Linux, (Mac) OS X and Windows. We click on download, follow the installation instructions and install R. https://www.r-project.org is the central website for the large community of R users. It is worth taking a look at this site. Here we find a lot of information about R, its development, the developers, manuals, a journal and the FAQs (Frequently Asked Questions).

After we have installed R, the R Console is at our disposal. R is now fully functional. However, when we open the R Console, we see that it appears very rudimentary and uninviting. We only see some information about R, e.g. that it is an open source software, and an input line where we can type in commands (Fig. 1.2: R Console).

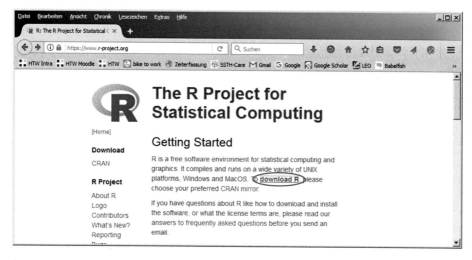

Fig. 1.1 r-project.org

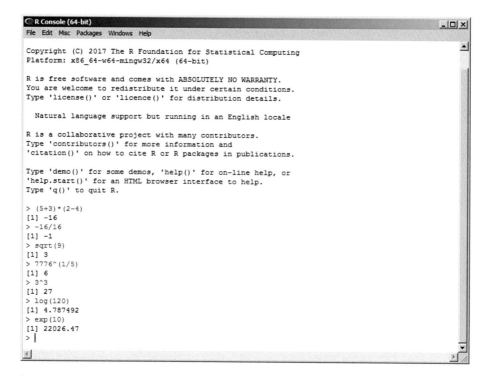

Fig. 1.2 R Console

Table 1.1 Some functions to calculate with in the R Console

Function	Description	Example
+	Addition of values	$5 + 3 = 8$
−	Subtraction of values	$2 − 4 = −2$
*	Multiplication of values	$8 * (−2) = −16$
/	Division of values	$−16/16 = −1$
sqrt()	Square root of a number	$sqrt(9) = 3$
(y)^(1 / x)	x-th root of the number y	$7776^\wedge(1/5) = 6$
^	Power of a number	$3^\wedge3 = 27$
log()	Natural logarithm	$log(120) = 4.79$
exp()	Exponential function	$exp(10) = 22,026.47$

Even if we do not normally use the Console directly for data analysis, as we said before, we can still use it very comfortably as a calculator on our computer. This is already a first introduction, which we warmly recommend.

In Table 1.1, we see some important calculator functions (also Fig. 1.2: R Console).

After installing R and maybe looking at the website and the console, we can start installing RStudio. The software for the different operating systems can be found on the website https://www.rstudio.com/ (Fig. 1.3).

Again we follow the link Download and get to a page where we can download several versions of RStudio, free desktop and server versions as well as paid desktop and server versions. We select the free desktop version, go to the installation files and take the installation file required for our operating system. We follow the installation instructions and install RStudio on our computer. If everything went smoothly, we can start RStudio now. Usually there are no problems, but if there are, we will look for help in the internet.

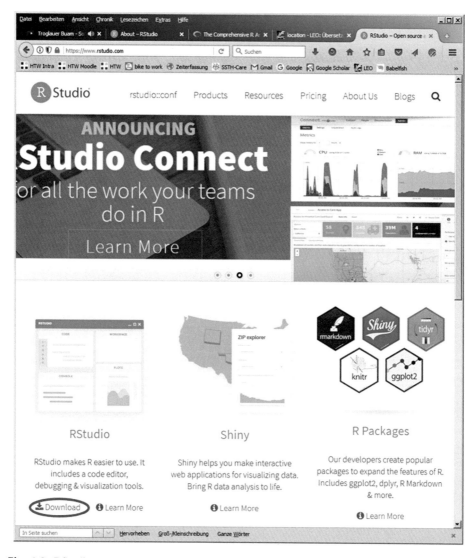

Fig. 1.3 RStudio

1.4   The Structure of RStudio

When you start RStudio, a four-parted screen opens, with the script window at the top left, the R Console at the bottom left, the data and object window at the top right, and the environment window at the bottom right. This is a good setup for data analysis. If RStudio does not appear in the view shown below after opening or if we want to have a different structure of RStudio, we can change the structure by going to "Tools" in the menu bar, selecting "Global Options" and making the desired settings under "Pane Layout".

At the top left, we find the *script window* (1). We use the script window to document our entire data analysis and store it permanently. Here we write down the finished commands, put them together, comment them and let them be processed one after the other. We save the finished script, so that we can always fall back on it and carry out the analysis again years later. In the script window, we write down the final copy of the entire analysis.

On the lower left, we see the *R Console* (2). We already got to know it and installed it with R. In the console, we see the results of our analysis when we execute the commands in the script window. We also have some space to try out commands. We can enter commands and see if they do what we want. This is an important function of the console in RStudio.

The *data and object window* (3) in the upper right corner shows us the data sets that are available in RStudio, i.e. loaded and active. Here we find not only the active datasets, but also other objects, e.g. vectors, functions that we generate during the analysis. We will go into this further later in the script.

In the lower right corner of the *environment window* (4) we find several tabs: Files, Plots, Packages, Help, etc. The Files tab shows the files stored in the active directory. In the *Plots* tab we see the generated graphics, in the *Packages* tab the installed and existing packages are listed and the *Help* tab we can use to request help on R and the existing functions. A little further down in the document we will deal with the help function in more detail.

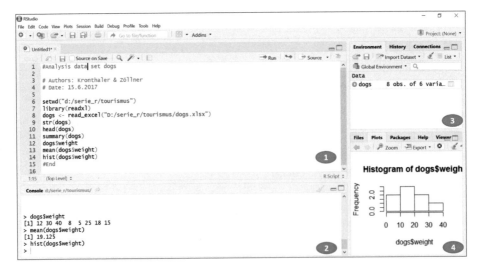

Fig. 1.4 Structure of RStudio

1.5   A First Data Analysis Application with RStudio

In the illustration above (Fig. 1.4: Structure of RStudio) we see a first small data analysis. Let's take a closer look at it to understand the logic of RStudio. How we create such a data analysis ourselves will be discussed in Chap. 2.

In the data and object window (3) we see that one data set is active. This data set has the name "dogs", contains 8 observations and 6 variables (8 obs[ervations] of 6 variables).

In the script window (1) we find the script of this short data analysis. Let's have a closer look at it (Fig. 1.5: Script dogs):

- In the first line we noted the headline of the data analysis, in the second line the authors of the script and in the third line the date when the script was created. Eye-catching in these lines is the hashmark #. The hashmark sign tells RStudio not to interpret the text as a command, but as a description/comment. We use the # whenever we insert a text into the script to comment on the data analysis.
- The fourth line contains the first command. *setwd("D:/serie_r/tourismus")* specifies the working directory of our computer (**set w**orking **d**irectory), which RStudio should use during the analysis and in which the results are stored. Setting the working directory at the beginning of the analysis is helpful, so we can easily find our results.

```
#Analysis data set dogs


# Authors: Kronthaler & Zöllner

# Date: 15.6.2017


setwd("d:/serie_r/tourismus")

library(readxl)

dogs <- read_excel("D:/serie_r/tourismus/dogs.xlsx")

str(dogs)

head(dogs)

summary(dogs)

dogs$weight

mean(dogs$weight)

hist(dogs$weight)

#End
```

Fig. 1.5 Script dogs

- The fifth line contains with *library(readxl)* a very important command for R. As we already mentioned, R is organized in packages for data analysis. The *library()* function tells R to use a specific package during the analysis. *readxl* is a package we need to read Excel files. In order to use the package, we need to tell RStudio that we need the package now. Usually we have to install packages once with the command *install.-packages()* before we can activate them every time we need them with the *library()* function. The next chapter shows how to install packages.
- The command *dogs <- read_excel("D:/serie_r/tourismus/dogs.xlsx")* tells RStudio to read the Excel file now. Let's take a closer look at the command. At the beginning, there is the name that the data set should get in RStudio. This is followed by the characters " <-", which is an assignment operator. It is followed by the actual command with which we import the Excel file, consisting of the command *read_excel*, the path and the name of the Excel file in parentheses and quotation marks. It is recommended to give the data set the same name the original data set has.
- The functions *str()*, *head()*, *hunde$gewicht* are functions to view the data. *summary()*, *mean()* are functions of data analysis and *hist()* is a function to create a graph. In the following we will see and deal with the functions more often.
- *# End* we finally use to mark the end of the script.

Now let's take a look at the R Console (2) and the environment window (4). Here we see the results of the analysis as well as the graphics. This reporting is not very comfortable. It is easier to save the results of the data analysis, respectively of the script, e.g. as PDF, as Word document or as HTML page. We do this using the key combination *Ctrl + Shift + K* or as described in the next figure using the relevant button (Fig. 1.6: Compile report). Note, before you can use the reporting function you have to save the script as .R file.

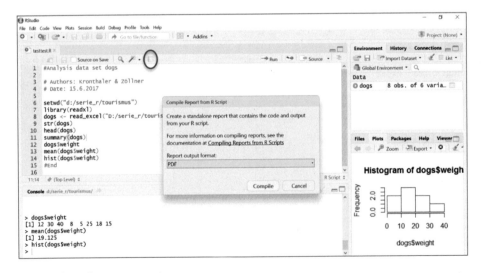

Fig. 1.6 Compile report

dogs.R

kronthafranz

Fri May 10 09:06:52 2019

```
#Analysis data set dogs

# Authors: Kronthaler & Zöllner
# Date: 15.6.2017

setwd("d:/serie_r/tourismus")
library(readxl)
dogs <- read_excel("D:/serie_r/tourismus/dogs.xlsx")
str(dogs)

## Classes 'tbl_df', 'tbl' and 'data.frame':    8 obs. of  6 variables:
## $ dog   : num  1 2 3 4 5 6 7 8
## $ sex   : num  0 1 1 1 0 1 1 0
## $ age   : num  5 10 12 20 7 2 1 9
## $ weight: num  12 30 40 8 5 25 18 15
## $ size  : chr  "medium" "big" "big" "small" ...
## $ breed : chr  "crossbreed" "dalmatiner" "shepherd" "terrier" ...

head(dogs)

## # A tibble: 6 x 6
##     dog   sex   age weight size   breed
##   <dbl> <dbl> <dbl>  <dbl> <chr>  <chr>
## 1     1     0     5     12 medium crossbreed
## 2     2     1    10     30 big    dalmatiner
## 3     3     1    12     40 big    shepherd
## 4     4     1    20      8 small  terrier
## 5     5     0     7      5 small  terrier
## 6     6     1     2     25 big    shepherd

summary(dogs)

##       dog            sex             age            weight
##  Min.   :1.00   Min.   :0.000   Min.   : 1.00   Min.   : 5.00
##  1st Qu.:2.75   1st Qu.:0.000   1st Qu.: 4.25   1st Qu.:11.00
##  Median :4.50   Median :1.000   Median : 8.00   Median :16.50
##  Mean   :4.50   Mean   :0.625   Mean   : 8.25   Mean   :19.12
##  3rd Qu.:6.25   3rd Qu.:1.000   3rd Qu.:10.50   3rd Qu.:26.25
##  Max.   :8.00   Max.   :1.000   Max.   :20.00   Max.   :40.00
##      size              breed
##  Length:8           Length:8
##  Class :character   Class :character
##  Mode  :character   Mode  :character
##
##
##

dogs$weight
```

Fig. 1.7 Report dogs

```
## [1] 12 30 40  8  5 25 18 15
mean(dogs$weight)
```

```
## [1] 19.125
hist(dogs$weight)
```

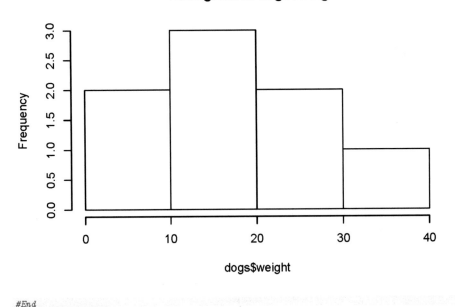

#End

Fig. 1.7 (continued)

If we click on the marked button or execute the key combination, we get the result report including commands (Fig. 1.7: Report dogs page 1 and page 2).

1.6   How to Install RStudio Packages

We have already mentioned several times that R or RStudio is based on packages with which we can carry out statistical analyses. In RStudio there are two ways to install and use packages. Packages can be installed menu-driven or via the corresponding command. The command is *install.packages( )*.

For example, the *readxl* package is required to read Excel files. To retrieve and install the corresponding package from the Internet we can enter the following command:

> install.packages("readxl").

It is important to have a working internet connection. *install.packages()* is the command, in brackets and quotation marks is the package to install. An important package for creating graphics is e.g. *ggplot2*. If we want to install it, we insert the name of the package into the command:

> install.packages("ggplot2").

The command *install.packages()* does not have to be entered via the script window. It is better to execute the command in the console. Usually we install a package only once and the command is therefore not part of our script.

Note that an installed package is not automatically active. To activate the package we use the *library()* command as mentioned above. Only with this command the package becomes active and can be used for data analysis. To activate *ggplot2* we enter the following command:

> library(ggplot2).

This command is usually part of our script (we'll see this again and again later).

If we use the menu-driven path, we go via the environment window, select the *Packages* tab and click *Install*. A window will open in which we can enter the package we want to install. RStudio will make suggestions as soon as we start typing the name. The following figure shows the menu-driven installation of the package *ggplot2*. Of course, we finish the installation with the button *Install* (Fig. 1.8).

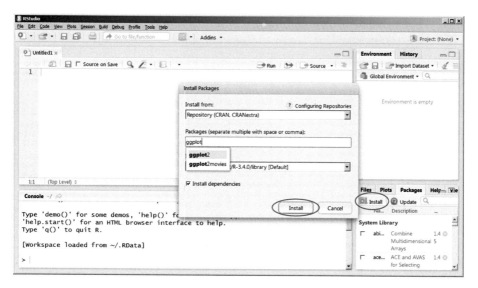

Fig. 1.8 Package installation with RStudio

But how do we know which package we need for which application? We don't worry about that for now. In the internet there is a lot of help for data analysis with R and RStudio. If we know which method we want to conduct, e.g. a factor analysis, then it is often enough to simply search for the corresponding keyword in combination with R or RStudio and we usually find a hint for a package.

Data Analysis Basics with RStudio

2

2.1 How to Read Data with RStudio

Before we start with the data analysis with RStudio, we first have to deal more intensively with reading data with RStudio. There is no data analysis without data. There are two ways of importing data into RStudio. We can use the *Import Dataset* button in the Data and Object window (Fig. 2.1: Import Data with RStudio) or we can enter and execute the appropriate syntax in the Script window. Different data formats can be imported into RStudio. Usually, data is available either as a text file, e.g. with the extension .csv, or as an Excel file with the extension .xlsx. We therefore discuss only these two data formats.

If we click the Import Dataset button, we can select the appropriate data format, e.g. *From Excel*. A window will open where we can search and import our Excel file. Depending on the format, we can also specify import options here. For an Excel file, for example, we can specify which sheet of the Excel workbook we want to import or that the first row contains the variable names. In this window, we can see the preview of the data set and are able to check visually whether the data is being read in correctly.

The second option is to directly enter the import command in the script window. We need to know the command and where the data file is located on our computer, i.e. the path. The command to import an Excel file is (if the readxl package is installed):

```
> library(readxl)
> dogs <- read_excel("D:/serie_r/tourismus/dogs.xlsx")
```

As already mentioned, we activate the *readxl* package with the *library()* command. The second step is to name the dataset. Usually we use the name of the Excel file to avoid confusion. The Excel file is called *dogs*. So we start the command with *dogs* followed by

© The Author(s), under exclusive license to Springer-Verlag GmbH, DE, part of Springer Nature 2021
F. Kronthaler and S. Zöllner, *Data Analysis with RStudio*,
https://doi.org/10.1007/978-3-662-62518-7_2

Fig. 2.1 Import data with RStudio

the assignment operator "<-". Then the command *read_excel()* follows, with the path and the Excel file in quotation marks inside the brackets.

With the path we must take into account that contrary to the normal procedure with the computer R does not work with backslashes, but with the forward slash "/".

To read the data we have to type in the command and send it. We can either proceed step by step or send the command at once. If we proceed step by step, we place the cursor on the respective line (of course we start with the *library()* command) and then click on the *Run* button in the upper right corner. If we proceed in one step, we mark all relevant lines with the mouse and send the commands again with the *Run* button (Fig. 2.2: Importing Excel data using the command syntax). Alternatively, we can use the key combination *Ctrl+Enter*.

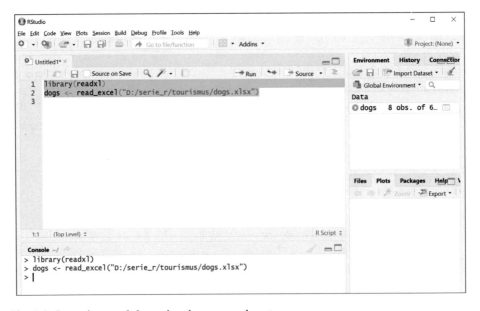

Fig. 2.2 Importing excel data using the command syntax

When the data import has been successfully completed, the (active) data set appears in the upper right window (Fig. 2.2: Importing Excel data using the command syntax).

If we want to import a CSV file instead of an Excel file, we do the same. If we work menu-driven, we select *From CSV* instead of *From Excel* and specify the necessary data. It is important that we know how the characters are separated from each other. For the CSV file it is usually the comma "," or the semicolon ";".

Practical tip

RStudio is sensitive to spelling and to upper and lower case letters. To simplify path and command input, it is recommended to avoid special characters, upper and lower case letters and to describe data sets and variables as simply as possible. In this text we follow this simple but helpful rule. So the path in which we store the files is kept relatively simple D:/serie_r/tourismus/, the data set is simply called dogs and the variable names dog, sex, age, weight, size and breed are kept short and are written in small letters.

2.2   How to Check Data with RStudio

The dataset is now loaded into RStudio. Before we start the data analysis, we should always check whether the data set has been read correctly. For this, we can use several commands. A first possibility is to use the function *head()* to display the top six rows of the data set, in other words the head of the data set. The command for this is:

```
> head(dogs)

## # A tibble: 6 x 6
##      dog        sex age    weight size     breed
##    <dbl>      <dbl> <dbl>    <dbl> <chr>    <chr>
## 1      1          0     5       12 medium   crossbreed
## 2      2          1    10       30 large    dalmatiner
## 3      3          1    12       40 large    shepherd
## 4      4          1    20        8 small    terrier
## 5      5          0     7        5 small    terrier
## 6      6          1     2       25 large    shepherd

# We get the first six rows of the record, including the column headers.
```

Alternatively, we can use the *tail()* function to display the last rows of the record. The command is as follows:

```
> tail(dogs)
```

```
## # A tibble: 6 x 6
##      dog       sex   age   weight size    breed
##      <dbl>   <dbl> <dbl>   <dbl> <chr>   <chr>
## 1       3         1    12      40 large   shepherd
## 2       4         1    20       8 small   terrier
## 3       5         0     7       5 small   terrier
## 4       6         1     2      25 large   shepherd
## 5       7         1     1      18 medium  crossbreed
## 6       8         0     9      15 medium  crossbreed
```

```
# We get the last six rows of the record, including the column headers.
```

With *View()* we can display the whole record in the script window:

```
> View(dogs)
```

To look at individual variables, we enter the name of the data set, followed by a dollar sign and the variable name. For example, if we only want to look at the variable weight, we enter the following command and see the weight of all dogs (compare the result of the command with the Fig. 2.3).

```
> dogs$weight
```

```
## [1] 12 30 40  8  5 25 18 15
```

Fig. 2.3 Have a look at the dataset with the View() command

Finally, let's take a look at the command *str()*. With this function, we get general information about the data set, about the variables, about the data type of the variables and about the characteristic values.

```
> str(dogs)
## Classes 'tbl_df', 'tbl' and 'data.frame':    8 obs. of  6 variables:
## $ dog      : num  1 2 3 4 5 6 7 8
## $ sex      : num  0 1 1 0 1 1 0
## $ age      : num  5 10 12 20 7 2 1 9
## $ weight   : num  12 30 40 8 5 25 18 15
## $ size     : chr  "medium" "medium" "large" "small" ...
## $ breed    : chr  "crossbreed" "dalmatiner" "shepherd" "terrier" ...
```

We see here for example the variables included in the data set, that the variable *age* is numeric (num) and that the variable *breed* is a character string (chr) respectively a text, with the expressions "crossbreed", "dalmatiner", etc.

2.3 Creating and Modifying Variables and Selecting Cases with RStudio

We have just seen that RStudio reads all variables whose expressions contain numbers as numeric values. Usually a data set is built with numbers, no matter if the variables are metric, ordinal or nominal. This means, once we have read in the data set, the first step is to prepare the data set accordingly. We can leave the metric variables as they are. The nominal and ordinal variables should be recoded into factors. In R, factors are nominal or ordinal data such as the classification of respondents by gender or educational level.

Using the example of the sex variable, we want to show how a new variable can be created with RStudio, how a variable is recoded to a factor, and how this new variable is added to the data set.

The command to convert an existing variable into a factor and simultaneously create a new variable is as follows.

```
> f_sex <- as.factor(dogs$sex)
```

f_sex is the name of the new variable. We ourselves typically rename an existing variable when converting it into a factor (but this is not necessary). "f" stands for factor.

as.factor is the command that converts the existing variable into a factor. After this, the name of the data set is given in parentheses, followed by the name of the variable, separated by a $ sign.

<- is the assignment operator that directs the variable we create to be stored under the name we specified.

Let's take a look at the data and object window on the top right, here the newly created variable *f_sex* has appeared. Now we have to add it to our dataset. The command is as follows:

```
> dogs <- cbind(dogs, f_sex)
```

The command is *cbind()*, "c" stands for column and bind stands for connect. We use this function to add a variable to an existing data set, more generally to connect vectors. In brackets, we first see the existing dataset and—separated by a comma—the variable we want to add.

The assignment operator "<-" and the name we specified before the assignment operator result in overwriting the existing dataset *dogs* and not creating a new dataset.

If we execute the command, the data and object window shows us that the dataset *dogs* now has seven instead of six variables.

Let's go back to our dataset *dogs*. We are able to display single rows or columns. For this, we need to know something about the logic of R. R first thinks in the rows of the data set, then in the columns. To select rows or columns we first enter the name of the data set, e.g. *dogs*, then we write the square bracket *dogs[]*. In the bracket, the rows are selected first and then, separated by a comma, the columns *dogs[rows, columns]*. We have a lot of possibilities. If we only want to display the first column and row, we enter the following command:

```
> dogs[1,1]
```

```
## [1] 1
```

If we want to display several rows and/or columns, we can define the range with a colon ":".

```
> dogs[1:4,1:3]
```

```
##      dog   sex   age
## 1     1     0     5
## 2     2     1    10
## 3     3     1    12
## 4     4     1    20
```

For example, if we want to see all rows, we can leave empty the first part before the comma (we get all rows with the defined columns):

```
> dogs[ ,1:3]

##    dog   sex   age
## 1    1    0     5
## 2    2    1    10
## 3    3    1    12
## 4    4    1    20
## 5    5    0     7
## 6    6    1     2
## 7    7    1     1
## 8    8    0     9
```

Instead of specifying numbers, we can also use names (usually, for a dataset, only the columns have names, so here we show this only for the columns). It is important that the column name is enclosed in quotation marks:

```
> dogs[ ,"sex"]

## [1] 0 1 1 1 0 1 1 0
```

If we want to display several rows or columns that are not necessarily one after each other, we need an addition in the command. The addition is c(), stands for combine, and with it the specified rows or columns are displayed. The following example gives us all rows and the columns *dog*, *sex* and *age*. Again, the column names are written in quotation marks and separated by a comma.

```
> dogs[ ,c("dog","sex","age")]

##     dog      sex   age
## 1    1        0     5
## 2    2        1    10
## 3    3        1    12
## 4    4        1    20
## 5    5        0     7
## 6    6        1     2
## 7    7        1     1
## 8    8        0     9
```

Here the interested reader can try to display rows 1, 3 and 4 for all columns of the data set dogs. Try it over the console.

Practical tip

The procedure just discussed can be used to create a reduced data set. This is sometimes convenient when only some of the variables or of the rows are included in the analysis. The following command creates a new data set with all rows and the first three columns

```
> dogs_reduced <- dogs[ ,1:3]
```

The name is dogs_reduced, followed by the assignment operator <- (see above) and the selected area of the data set.

One last hint: If we are unsure how a variable is defined in RStudio, we can execute the *class()* function. It gives information about the type of the variable. The most important types are:

- numeric for metric variables,
- factor for nominal or ordinal variables,
- character for text.

For example, we are interested in the type of variable *age* of the dataset *dogs*. The following command gives information about it:

```
> class(dogs$age)

## [1] "numeric"

# The variable *age* is numeric.
```

During the preparation and analysis of the data, one often faces the task of changing variables or generating new variables. Basically one can use the available arithmetic operations (see calculator functions above, of course there are not only these operations). Here, we want to show only briefly and with a small example how a new variable is generated. Let's imagine that we want to convert dog age into human age and produce the new variable *humanage* for this purpose.

The command has the same structure as a variable converted into a factor has. We first define the variable name of the new variable, then write the assignment operator and then the calculation rule. The command looks like this if we assume that a dog year corresponds to six human years, i.e. we multiply the variable *age* by six.

```
> humanage <- dogs$age*6
```

With the *cbind()* function we only have to add the new variable to the data set. Then it is available in the data set.

```
> dogs <- cbind(dogs, humanage)
```

2.4 Commands and Command Structure in RStudio

We have already learned some commands of R and RStudio. Let's take a closer look at them.

There are some commands that are used repeatedly. These are the following commands (Table 2.1).

If we look at the structure of the commands, we see that we write the command first, followed by an open and closed round bracket. Within the parenthesis, we specify what and how the command should be applied to.

ls() is an exception, the function always lists all variables and objects active in the working memory. The round parenthesis always remains empty.

```
> ls()

## [1] "f_sex"  "dogs"       "dogs_reduced"     "humanage"

# We receive the currently four active data sets and variables "dogs",
"dog_reduced", "f_sex" and "humanage".
```

rm() deletes single specified variables and objects. For example, if we no longer need the dataset *dogs_reduced* and the variable *f_sex*, we enter the following command (we separate the variables/objects with a comma):

```
> rm(dogs_reduced, f_sex)
# The dataset dogs_reduced and the variable f_sex are now removed from the
working memory.
```

With *rm(list=ls())* without further specifications we clear the whole memory.

```
> rm(list=ls())
# The working memory is cleared.
```

Table 2.1 Commands used repeatedly

Command	Description
ls()	Lists all active variables and objects within the working memory
rm()	Deletes selected variables and objects within the working memory
rm(list=ls())	Deletes all variables and objects within the working memory
str()	Displays important information about the structure of the dataset
View()	Displays the dataset in a new window
help()	Gives help for commands

The working memory is now empty and we can start all over again. If we look at the data and object window at the top right, we see it is now without content.

Before we can go on, we have to reload our dataset again:

```
> library(readxl)
> dogs <- read_excel("D:/serie_r/tourismus/dogs.xlsx")
```

str() gives us important information of the dataset, for this we have to specify the dataset within the round brackets:

```
> str(dogs)

## Classes 'tbl_df', 'tbl' and 'data.frame':    8 obs. of  6 variables:
## $ dog      : num  1 2 3 4 5 6 7 8
## $ sex      : num  0 1 1 1 0 1 1 0
## $ age      : num  5 10 12 20 7 2 1 9
## $ weight   : num  12 30 40 8 5 25 18 15
## $ size     : chr  "medium" "medium" "large" "small" ...
## $ breed    : chr  "crossbreed" "dalmatiner" "shepherd" "terrier" ...
```

We see that the data set contains 8 observations and 6 variables, in addition we see the description of the variables.

We use *View()*, as shown above, if we want to view the entire dataset in a separate window. Also here we have to specify the name of the dataset:

```
> View(dogs)
```

With *help()* we call for help about functions, in the brackets we specify the function for which we want to get help. For example, if we want to know more about the mean function, we enter the following command:

```
> help(mean)

## starting httpd help server ... done

# In our Internet browser, a window opens with information about the command.
```

We see that commands are executed in R by typing the name of the command followed by a parenthesis. In the parentheses, we usually start with specifying the dataset and the variable for which we want to perform the calculation. The name of the dataset comes first, followed by a dollar sign *$* and the name of the variable. For example, if we want to know the number of male and female dogs, we can use the *table()* function. By default, this function gives us a table with the number of observations per variable characteristic. The command is as follows:

```
> table(dogs$sex)

##
## 0 1
## 3 5
```

We see that the data set contains three observations with the characteristic 0 (male) and five observations with the characteristic 1 (female).

If we additionally want to tabulate the number of male and female dogs according to size, the command changes a little bit:

```
> table(dogs[ ,c("sex", "size")])

##                 size
##       sex   large small medium
##       0       0     1      2
##       1       3     1      1
```

table() is the function for counting and tabulating values. In the parenthesis is the dataset defined, followed by a square bracket, with which we make the selection within the data set. Here we have to remember the logic of R again (see above). First, we select the rows. Since we want to consider all rows of the dataset, we leave the part before the comma empty. After the comma we specify the variables (columns) with *c()*. We write them in quotation marks, separated by a comma. With the function *table()* we get the counted observations for the sex according to the size of the dogs.

Now we know about some general features that help to understand the command structure of R. In the following, we will encounter these features repeatedly and we will slowly incorporate them.

Now we have already done a lot and would like to secure our work. How this is to be accomplished best is discussed in the following chapter.

2.5 Script Files and Reporting

When saving the work, we can distinguish between saving the work steps and saving the work results. The former we need if we want to carry out the analysis again at a later point, the latter we need in order to save the results of the data analysis permanently and if necessary process them in a text.

As already shown in Sect. 1.4, the steps are saved using a script file. Now we want to deal with the creation of a script file more detailed.

We start with an empty script file. To create one we go to *New File* in the *File* tab and then click on *R Script*. Alternatively we can press the key combination *Ctrl+Shift +N* (Fig. 2.4: Create a R script file).

Now we can start to write the script and design it in a meaningful way (Fig. 2.5: Example script).

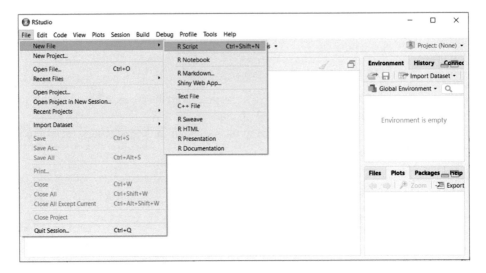

Fig. 2.4 Create a R script file

We usually start with the title of the data analysis, the name of the creator of the script and the date the script was written (lines 3, 6 and 7 in the following figure). The hashmark # tells R that there is text in these lines and nothing to calculate. The dotted lines above and below the title (lines 2 and 4) are not necessary. In our case, they only serve as an optical separation to emphasize the title. At the same time, you create chapters by using a hashmark followed by the dotted lines. On the left side next to the numbering of the lines, there is a small triangle. You can open and close the subchapters by clicking on the triangle.

With the first section of the script, we prepare R for the data analysis. We clean up our desk and empty the memory with the command *rm(list=ls())*. It makes sense to delete all objects stored in R before creating the script. This prevents confusion with other datasets. Additionally we define our working directory with *setwd()*, i.e. the place where R searches for data and stores data and results.

In the second and third sections, we read in the data, check it and prepare it. To read in the data we use the package *readxl* and the function *read_excel()*. To control the data we can use the command *View()*. The Data preparation section is relatively short, and contains only the commands for creating a factor. Under certain circumstances, this section can become relatively long, for example, if we need to delete values or create new variables, and so on.

The fourth section starts with the data analysis. Depending on the question and task, this area naturally varies in length. However, it will always begin with a descriptive part and then switch to the inductive statistics. It is important to document each data analysis step clearly, so the script remains manageable and it becomes clear what is done with the commands.

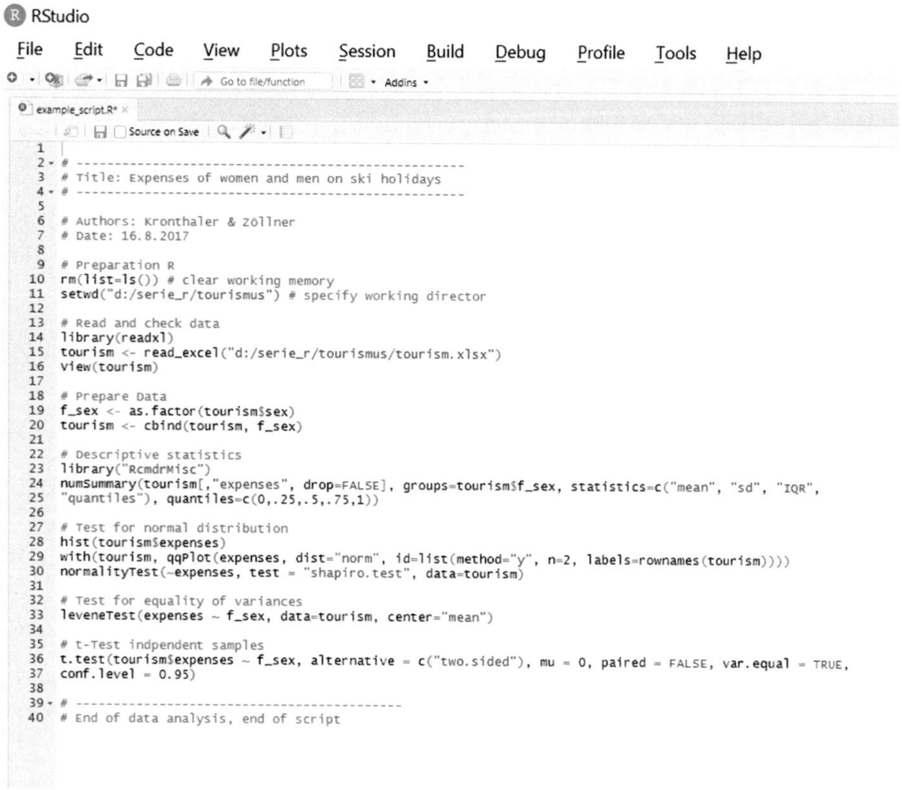

Fig. 2.5 Example script

Finally, we end the script with the information that the data analysis is finished and the script ends.

This short example sufficiently illustrates how a script is structured. In reality, the analyses are of course longer, but the described procedure is always the same and so is the structure of the script. For longer data analyses, we can also divide the script into sub-chapters, so that the script remains straightforward.

When the script is ready, we should save it. It probably makes sense to do this step at the beginning of the script creation and to save it repeatedly in case the computer crashes. To save the script we can either go back to the menu, click on *File* in the upper left corner and then on *Save* or use the key combination *Ctrl+S*. Of course, we give the script a suitable name and use the extension .R *scriptname.R*. Once we have saved the script, we can close it and open it again if necessary. To open a script we use the menu, go to *File*, then to *Open File* and look for our script with the extension .R. Now we can edit the script or do the analysis again.

Fig. 2.6 Reporting with RStudio

The script contains only the documentation of the work steps and the calculation steps carried out. We also need the results of the work. To get the results, we use the reporting function in R when the script is open. It is called *Knit Document* and is located in the *File* tab, alternatively we can use the key combination *Ctrl+Shift+K* or click on the *Compile Report* button (Fig. 2.6: Reporting with RStudio).

If we click on the Compile Report button, a window opens with the reporting options. We can create an HTML page with the results or a PDF or a Word document. Depending on the preference and the question of further processing, we will select one or the other. If we further want to process the results in a Word document, we certainly recommend the Word document. If we only want to save the results permanently, we recommend the HTML document or the PDF. If, for example, we select the PDF, we receive the results, including the executed commands, as PDF (Fig. 2.7: Save working results).

The report is not completely listed here, only the beginning.

Now we know the basics of data analysis with RStudio and can focus on data analysis. For this purpose, we use a simulated data set, which is described in Chap. 3.

example_script.R

kronthafranz

Fri May 10 16:40:03 2019

```
# Title: Expenses of women and men on ski holidays

# Authors: Kronthaler & Zöllner
# Date: 16.8.2017

# Preparation R
rm(list=ls()) # clear working memory
setwd("d:/serie_r/tourismus") # specify working director

# Read and check data
library(readxl)
tourism <- read_excel("d:/serie_r/tourismus/tourism.xlsx")
View(tourism)

# Prepare Data
f_sex <- as.factor(tourism$sex)
tourism <- cbind(tourism, f_sex)

# Descriptiv statistics
library("RcmdrMisc")
```

```
## Loading required package: car
```

```
## Loading required package: carData
```

```
## Loading required package: sandwich
```

```
numSummary(tourism[,"expenses", drop=FALSE], groups=tourism$f_sex, statistics=c("mean", "sd", "IQR",
"quantiles"), quantiles=c(0,.25,.5,.75,1))
```

```
##        mean       sd IQR  0% 25% 50% 75% 100% expenses:n
## 0 349.3647 46.39552  68 214 315 352 383  477         85
## 1 392.0462 43.79406  76 284 353 401 429  481         65
# Test for normal distribution
hist(tourism$expenses)
```

Fig. 2.7 Save working results

Table 2.2 R commands learned in Chap. 1 and in this chapter

Command	Notes/examples
as.factor()	....
cbind()	....
class()	....
head()	....
help()	....
install.packages()	....

(continued)

Table 2.2 (continued)

Command	Notes/examples
library()	....
ls()	....
read_excel()	....
rm()	....
rm(list = ls())	....
setwd()	....
str()	....
table()	....
tail()	....
View()	....

Note You can take notes or write down examples for their usage on the right side

2.6   Time to Try

2.1. Use the R Console as calculator. Calculate the following terms: $(5 + 2)*(10 - 3)$, square root of 64, third root of 64, third root of 512, 4 to the third power, 8 to the third power, logarithm of 500, logarithm of 10, Euler's number exp(1), exponential value e^x of x = 5.

2.2. Download the dataset *dogs.xlsx* from http://www.statistik-kronthaler.ch/ and save the dataset in a folder on your laptop. Try to use a short path and a folder with a simple name. Remember the path and the folder; it is the way and the place to find the data.

2.3. Read the dataset *dogs.xlsx* via Import Dataset button. Delete the dataset with the command *rm()*. Read the dataset again with the import command in the script window. Delete your working memory completely with the correct command.

2.4. Read the dataset *dogs.xlsx* and try to do the following tasks:

- Have a look at the dataset with the relevant command. Check whether the data set is complete by looking at the beginning and at the end of the dataset with the relevant commands.

- Have a look at the general information about the data set, i.e. the data type of variables, characteristic values, and so on.

2.5. Produce a reduced dataset with only 6 observations and without the variable breed. Name the dataset *dogs_red*.

2.6. Work with the dataset *dogs_red* produced in task 5. How old is the dog in the fourth observation? Try to figure it out by displaying only this value. Check the value received with the help of the command *View()*.

2.7. Generate with *humanage* one new variable that is added to the dataset *dogs.xlsx*. You heard that one dogyear is only 5 humanyears not 6. Furthermore, convert the variables *sex*, *size* and *breed* into factors.

2.8. Produce your first small script with the help of the dataset *dogs.xlsx* (you can do the same with the reduced dataset produced in task 5), save it and save the reporting as word file as well. To do so include all the relevant issues and use for the data analysis the following commands (we will learn about the commands in Chap. 4):

- summary(dogs);
- hist(dogs$age);
- numSummary(dogs[,"age", drop=FALSE], statistics=c("mean", "sd", "IQR", "quantiles"), quantiles=c(0,.25,.5,.75,1));
- boxplot(dogs$age ~ dogs$f_sex);
- numSummary(dogs[,"age", drop=FALSE], groups=dogs$f_sex, statistics=c ("mean", "sd", "IQR", "quantiles"), quantiles=c(0,.25,.5,.75,1)).

Data Tourism (Simulated)

3

In order to learn data analysis with RStudio, we obviously need to use data. We have two types of data: secondary data and primary data. Secondary data is data that someone else has collected and made available for data analysis. Primary data is data that we collect ourselves to answer our questions.

There is a lot of secondary data that is available publicly and that we can use. Sources for secondary data include the Statistical Offices of countries, the Statistical Office of the European Union, the OECD, the World Bank and the IMF. For tourism in Switzerland for example, the statistics HESTA of the Swiss Federal Statistical Office BfS is of particular interest. Here we find numbers of guests and overnight stays by region and country of origin, the number of hotels, rooms and beds. These data are available annually. For other countries, e.g. Germany and Austria, similar figures are certainly available. It is always worth searching for secondary data.

In some cases, however, there is no secondary data and we are forced to collect our own (primary) data. In this text we follow this approach. In this way, we believe we can demonstrate the benefits of RStudio and provide a simple and intuitive approach to data analysis with RStudio.

Let us assume that the mountain railways of the *Heide* would like to adjust their offering to the wishes and needs of their guests in order to improve the attractiveness of the ski area. For this purpose the management carries out a representative short survey and asks 150 guests (n = 150) about their satisfaction with the current offering (see questionnaire in Appendix A).

After the survey has been carried out, the data must be stored in a file format. If we have carried out the survey with the help of tablets and standard survey software (Unipark, LimeSurvey, etc.), the data is already available electronically and we can export the dataset e.g. as CSV file or Excel file. If we have carried out the survey with paper questionnaires, the first step is to enter the data. Experience shows that the data should be entered directly into an Excel worksheet, it is the easiest way.

© The Author(s), under exclusive license to Springer-Verlag GmbH, DE, part of Springer Nature 2021
F. Kronthaler and S. Zöllner, *Data Analysis with RStudio*,
https://doi.org/10.1007/978-3-662-62518-7_3

In Appendix B we see the dataset as it might look like. We also see that we have created a legend for the data set that describes the dataset. Alternatively, we could create a so-called codebook.

> *Note!*
>
> **At this point, a note is important: The questionnaire was not really carried out and was only created for this text. The data are all simulated data.**

The dataset is called tourism.xlsx and we first import it into RStudio. Typically, we delete our memory first to have a clean working space.

```
> rm(list=ls())
> library(readxl)
> tourism <- read_excel("D:/serie_r/tourismus/tourism.xlsx")
```

Then we check the data set to see if it has been read correctly. We can do this with the commands *str()*, *head()*, *tail()* and *View()*. At the same time, we get a first idea of the data.

The command *str()* gives us a good overview of our dataset. We see that the dataset contains 150 observations and 16 variables. We also see the names of the variables and the first ten values of the variables. In addition, we see that RStudio has read all variables in number format (num[eric]). This is because the variables contain only numbers. If we look at the legend in Appendix B, we see that the data set contains not only metric variables, but also nominal and ordinal variables. We have to tell RStudio about this and prepare the data set accordingly.

```
> str(tourism)

## Classes 'tbl_df', 'tbl' and 'data.frame':    150 obs. of  16 variables:
##  $ guest          : num  1 2 3 4 5 6 7 8 9 10 ...
##  $ accommodation  : num  3 4 3 2 2 1 3 3 3 3 ...
##  $ stay           : num  5 5 7 5 2 7 6 6 5 8 ...
##  $ diversity      : num  41 90 78 84 68 77 98 48 100 96 ...
##  $ waitingtime    : num  31 68 43 44 33 39 57 61 91 73 ...
##  $ safety         : num  91 76 76 61 76 94 68 90 78 100 ...
##  $ quality        : num  25 73 10 26 21 55 22 80 18 10 ...
##  $ satisfaction   : num  67 63 49 64 48 79 63 62 96 81 ...
##  $ price          : num  1 3 1 1 2 3 2 2 3 3 ...
##  $ expenses       : num  368 427 331 341 347 359 351 383 444 394 ...
##  $ recommendation : num  1 3 4 3 3 2 3 3 1 1 ...
##  $ skiholiday     : num  1 1 0 0 1 0 1 1 1 0 ...
##  $ sex            : num  0 1 0 1 0 0 1 0 1 0 ...
##  $ country        : num  1 1 2 4 2 2 1 3 1 1 ...
##  $ age            : num  42 50 44 41 43 38 47 66 62 49 ...
##  $ education      : num  4 3 2 1 4 2 3 4 1 4 ...
```

The *head()* function reads the first six rows of the dataset.

```
> head(tourism)

## # A tibble: 6 x 16
##    guest accommodation  stay diversity   waitingtime   safety
quality
##    <dbl>       <dbl> <dbl>   <dbl>       <dbl>     <dbl>    <dbl>
## 1     1           3     5    41.0          31        91       25
## 2     2           4     5    90            68        76       73
## 3     3           3     7    78            43        76       10
## 4     4           2     5    84            44        61       26
## 5     5           2     2    68            33        76       21
## 6     6           1     7    77            39        94       55
## # ... with 9 more variables: satisfaction <dbl>, price <dbl>,
## #   expenses <dbl>, recommendation <dbl>, skiholidy <dbl>, sex <dbl>,
## #   country <dbl>, age <dbl>, education <dbl>
```

The *tail()* function shows the last six rows of the data set.

```
> tail(tourism)
```

```
## # A tibble: 6 x 16
##     guest accommodation      stay diversity waitingtime      safety
quality
##     <dbl>         <dbl>     <dbl>     <dbl>       <dbl>       <dbl>     <dbl>
## 1     145             4         7      87.0          60        97.0        46.0
## 2     146             2         9        67         44.         83.          57
## 3     147             4         6        66          55          96          40
## 4     148             3         5        80          69          79          38
## 5     149             3         8        60          40          68          43
## 6     150             3         5        55          15          61          31
## # ... with 9 more variables: satisfaction <dbl>, price <dbl>,
## #    expenses <dbl>, recommendation <dbl>, skiholiday <dbl>, sex <dbl>,
## #    country <dbl>, age <dbl>, education <dbl>
```

With *View()* we display the dataset in the script window (Fig. 3.1: Viewing the dataset in the R window).

```
> View (tourism)
```

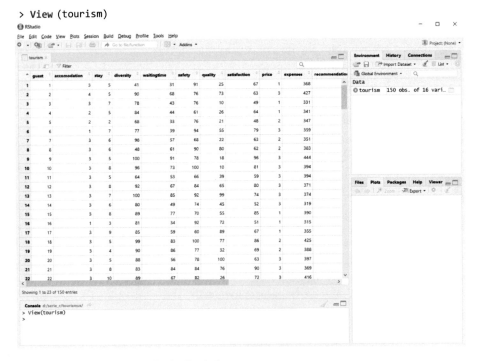

Fig. 3.1 Viewing the dataset in the R window

After we have checked the data and are familiar with the data set, we can start with the data analysis.

Describing Data with RStudio

4

One of the important tasks in the process of data analysis is getting to know and describing the data set with the help of key figures and graphics.

Before we start, we first delete our memory and reload the dataset.

```
> rm(list=ls())
> library(readxl())
> tourism <- read_excel("D:/serie_r/tourismus/tourism.xlsx")
```

4.1 Descriptive Key Figures

Important key figures of a dataset are mean values, measures of variation and correlation coefficients. Every article in your field of study that works with data will discuss these key figures.

Frequently used mean values are the arithmetic mean, the median and the mode. The arithmetic mean indicates the average of a variable by first adding up all values and then dividing them by the number of observations. The median is the value that divides the values of a variable into the 50% smallest and the 50% largest values. The mode is the value that occurs most frequently.

Important measures of variation are the range, the quartile deviation, the standard deviation and the variance. The range is the difference between the smallest and the largest value within a variable. The quartile deviation is the difference between the 1st quartile (25% of the values are smaller than the 1st quartile) and the 3rd quartile (75% of the values are smaller than the 3rd quartile) of a variable. The 50% of the middle values are within the quartile deviation. The standard deviation can be interpreted as the average deviation of the individual values from the mean value and the variance is the squared standard deviation.

F. Kronthaler and S. Zöllner, *Data Analysis with RStudio*, https://doi.org/10.1007/978-3-662-62518-7_4

Correlation coefficients often used in practice are the Bravais-Pearson correlation coefficient for metric variables, the Spearman correlation coefficient for ordinal variables, and the contingency coefficient for nominal variables.

We can display many of these values using the *summary()* command.

```
> summary(tourism)

##       guest         accommodation        stay            diversity
## Min.   :  1.00    Min.   :1.000     Min.   : 1.000    Min.   :  0.00
## 1st Qu.: 38.25    1st Qu.:3.000     1st Qu.: 5.000    1st Qu.: 64.00
## Median : 75.50    Median :3.000     Median : 6.000    Median : 78.00
## Mean   : 75.50    Mean   :2.987     Mean   : 6.047    Mean   : 75.62
## 3rd Qu.:112.75    3rd Qu.:3.000     3rd Qu.: 7.000    3rd Qu.: 90.75
## Max.   :150.00    Max.   :5.000     Max.   :13.000    Max.   :100.00
##   waitingtime        safety           quality         satisfaction
## Min.   : 3.00     Min.   : 52.00    Min.   :  0.00    Min.   :15.00
## 1st Qu.:42.25     1st Qu.: 69.00    1st Qu.: 32.25    1st Qu.:53.25
## Median :55.00     Median : 79.50    Median : 48.50    Median :65.50
## Mean   :54.70     Mean   : 79.72    Mean   : 48.84    Mean   :63.64
## 3rd Qu.:69.00     3rd Qu.: 90.00    3rd Qu.: 64.00    3rd Qu.:76.75
## Max.   :95.00     Max.   :100.00    Max.   :100.00    Max.   :98.00
##     price            expenses       recommendation     skiholiday
## Min.   :1.000     Min.   :214.0     Min.   :1.000     Min.   :0.00
## 1st Qu.:1.000     1st Qu.:334.2     1st Qu.:2.000     1st Qu.:0.00
## Median :2.000     Median :368.5     Median :2.000     Median :1.00
## Mean   :2.133     Mean   :367.9     Mean   :2.513     Mean   :0.62
## 3rd Qu.:3.000     3rd Qu.:405.8     3rd Qu.:3.000     3rd Qu.:1.00
## Max.   :3.000     Max.   :481.0     Max.   :4.000     Max.   :1.00
##     sex             country          age             education
## Min.   :0.0000    Min.   :1.000     Min.   :16.00    Min.   :1.00
## 1st Qu.:0.0000    1st Qu.:1.000     1st Qu.:41.00    1st Qu.:1.00
## Median :0.0000    Median :2.000     Median :47.50    Median :2.00
## Mean   :0.4333    Mean   :1.907     Mean   :47.91    Mean   :2.18
## 3rd Qu.:1.0000    3rd Qu.:2.750     3rd Qu.:57.00    3rd Qu.:3.00
## Max.   :1.0000    Max.   :4.000     Max.   :79.00    Max.   :4.00
```

For each variable, the command specifies the smallest value, the 1st quartile, the median, the arithmetic mean, the 3rd quartile and the largest value. The range and the quartile deviation can be easily calculated from this. The central tendency can be determined by the arithmetic mean and the median. We learn an important lesson here, R, like any other statistical software, calculates what we specify, whether it makes sense or not. For example, the variable country is a nominal variable, neither the 1st quartile, nor the median, the arithmetic mean, or the 3rd quartile can be meaningfully calculated and interpreted.

Alternatively, we can also select the individual measures for individual variables. With the following command, we can figure out the mode for the variable *country*.

```
> table(tourism$country)

##
##  1  2  3  4
## 64 48 26 12
```

The *table()* function tells R to count the number of occurrences of variable values. Here we see that the number one most often occurs, i.e. the mode is one or Switzerland (see legend of the data set).

We display the median with *median()*.

```
> median(tourism$safety)

## [1] 79.5
```

The median for the variable *safety* is 79.5, i.e. 50% of the values are smaller than 79.5 and 50% of the values are larger.

The arithmetic mean value is obtained with the *mean()* function.

```
> mean(tourism$safety)

## [1] 79.72
```

Measured with the arithmetic mean, the average satisfaction with safety on the slope is 79.72 points (see legend of data set).

The functions *quantile()*, *sd()* and *var()* are available for the measures of variation.

The *quantile()* displays by default the 0% quantile (smallest value), the 1st quartile (25% percentile), the 2nd quartile (50% percentile), the 3rd quartile (75% percentile) and the 100% quantile (largest value).

```
> quantile(tourism$safety)

##    0%   25%   50%   75%  100%
## 52.0  69.0  79.5  90.0 100.0
```

Considering the example, we see that the smallest value is 52.0; 25% of the values are smaller than 69.0; 50% of the values are smaller than 79.5; 75% of the values are smaller than 90.0 and the largest value is 100.0. Note, the 2nd quartile is the median.

With *sd()* we get the standard deviation. It is about 13.13 points.

```
> sd(tourism$safety)
```

```
## [1] 13.12773
```

var() displays the variance. It is the squared standard deviation and it is in our case 172.34. We can try to square the standard deviation in the console. We should get the same figure except for rounding errors.

```
> var(tourism$safety)
```

```
## [1] 172.3372
```

It is often convenient to obtain several key figures at the same time for individual variables. Various packages are available for this purpose. One possibility is to activate the package *RcmdrMisc* and then work with the function *numSummary()*.

```
> library("RcmdrMisc")
## Loading required package: car

## Loading required package: carData

## Loading required package: sandwich
```

```
> numSummary(tourism[,c("waitingtime", "safety"), drop=FALSE],
statistics=c("mean", "sd", "IQR", "quantiles"), quantiles=c(0,.25,.5,.75,1))
```

```
##                 mean       sd   IQR 0%   25% 50% 75% 100%    n
## waitingtime    54.70 20.98569 26.75  3 42.25 55.0  69   95 150
## safety         79.72 13.12773 21.00 52 69.00 79.5  90  100 150
```

Before we go to the result, let us briefly look at the command. With *numSummary()* we instruct R to give us descriptive statistics. The open parenthesis is followed by the dataset. With the square bracket, we specify the variables for which we want to have the key figures. Then we specify the key figures. The way the command is formulated, we get the arithmetic mean value, the median (2nd quartile), the interquartile range as well as the 0%, 25%, 50%, 75%, 100% quantiles for the variables *waiting time* and *safety* of the dataset *tourism*.

The average satisfaction measured by the arithmetic mean and the median for the variable *waiting time* is much lower than for the variable *safety*. In addition, we see, measured both by the standard deviation and the interquartile range, that the assessment of the guests for the variable *waiting time* is much more heterogeneous, the variation is larger.

Frequently we want to see the descriptive statistics not only for the variable as a whole, but for certain groups. Here, too, the *RcmdrMisc* package just activated is helpful. Let's imagine, we want to know how men and women feel about safety and want to have a look at the descriptive statistics. To analyze this, we first have to define the variable gender as a factor:

```
> f_sex <- as.factor(tourism$sex)
> tourism <- cbind(tourism, f_sex)
```

Now we are able to extend the *numSummary()* function. Therefore, we insert the procedure *groups=tourism$f_sex* in the middle of the function:

```
> library("RcmdrMisc")
> numSummary(tourism[,"safety", drop=FALSE], groups=tourism$f_sex,
statistics=c("mean", "sd", "IQR", "quantiles"), quantiles=c(0,.25,.5,.75,1))

##        mean        sd IQR 0% 25% 50% 75% 100% sicherheit:n
## 0 80.88235 12.66173  21 52  70  84  91  100           85
## 1 78.20000 13.66245  23 52  67  78  90  100           65
```

The result displays the statistical figures for men and women. We see that men are slightly more satisfied than women, both in terms of the arithmetic mean and the median (0 is the coding for men and 1 for women, see legend). We also see, measured by the standard deviation and the inter quartile range, that the variation is approximately the same.

Finally, let's have a look at the commands to display correlation coefficients. We differentiate between a correlation matrix for many variables and a correlation coefficient for a pair of variables.

The correlation matrix is used in particular for a larger number of metric variables. The command for a correlation matrix with several variables is as follows:

```
> cor(tourism[,c("quality","safety","diversity","waitingtime",
"satisfaction")], method="pearson", use="complete")

##                   quality      safety    diversity waitingtime satisfaction
## quality       1.000000000  0.02820157  0.11103043  0.08553185  0.007709465
## safety        0.028201573  1.00000000 -0.09081689  0.01849999 -0.022505599
## diversity     0.111030428 -0.09081689  1.00000000  0.76150583  0.613930271
## waitingtime   0.085531849  0.01849999  0.76150583  1.00000000  0.793096420
## satisfaction  0.007709465 -0.02250560  0.61393027  0.79309642  1.000000000
```

With *cor()* we specify that we want to see correlation coefficients, these for the dataset tourism and for the variables *quality*, *safety*, *diversity*, *waitingtime* and *satisfaction*. With *method="pearson"* we tell R to calculate the correlation coefficient of Bravais-Pearson and with *use="complete"* we tell R to use only those lines in which all values are available. In the square bracket, we address the rows and columns of the data set as mentioned above (see Sect. 2.4). If nothing is specified before the comma we use all rows in the data set, the variables we want to include in the correlation matrix are specified after the comma.

To interpret the correlation matrix, we only need to know that the correlation coefficient of Bravais-Pearson is defined between -1 and 1, where -1 describes a perfectly negative relationship and 1 a perfectly positive relationship. Zero means no relationship between two variables. For example, we see that most correlation coefficients are close to zero, but that we have stronger positive relationships between the variables *diversity* and *waitingtime*, *diversity* and *satisfaction* and *waitingtime* and *satisfaction*.

If we have ordinal variables instead of metric variables and want to display the correlation matrix, we simply specify the command with *method="spearman"* instead of *method="pearson"*. The interpretation is similar to the interpretation of the correlation coefficient of Bravais-Pearson.

If we want to display the correlation coefficients for a pair of variables, we also use the function *cor()*. Below we see the command for the correlation coefficient of Bravais-Pearson for metric data and the correlation coefficient of Spearman for ordinal data. The difference lies in whether we select *method="pearson"* or *method="spearman"* as shown above.

```
> cor(tourism$diversity, tourism$waitingtime, method="pearson",
use="complete")

## [1] 0.7615058


> cor(tourism$accommodation, tourism$education, method="spearman",
use="complete")

## [1] -0.01094637
```

The contingency coefficient, the correlation coefficient for nominal variables, is not discussed here. We refer to it in Chap. 6 when discussing how to test for a relationship between nominal variables.

The commands shown allow us to calculate the descriptive statistics for the data set as a whole, for individual variables and for groups. They include the most important applications for descriptive statistics.

4.2 Statistical Charts

We have now seen how RStudio can be used to obtain important descriptive statistical key figures. The visual display of the data is just as important. Statistical graphs show us patterns, relationships, unusual facts and observations. Graphics also create a lasting visual impression. Therefore, it always makes sense to check the data graphically for patterns, relationships and unusual observations.

With this, we already have indirectly noted that we need graphs for two reasons. First, we need graphs during data analysis to analyze existing patterns and relationships, as well as to check for unusual observations. Second, we need graphs to visualize and present facts. For the former, it is not very important how the graphs are designed; they are only used for analysis purposes. For the second, the visual design is important as well. If we present graphs or would like to publish them, then the appearance of the graphs is decisive for their credibility.

In the following, we proceed in two steps. First, we show how RStudio can be used to generate important statistical graphs quickly and easily. Then we discuss how to create advanced graphs with a very popular graphics package called *ggplot2*.

Important Statistical Charts While Analyzing Data

Simple graphs that are always needed during data analysis are the histogram, the pie chart, the bar chart, the box plot and the scatter plot. The commands to display these graphs are: *hist()*, *pie()*, *barplot()*, *boxplot()* and *plot()*. Often the default setting is sufficient and only the required variables need to be specified. Sometimes it is useful to label the axes and give the graph a title during data analysis. For this purpose, we can add various extensions to the commands, e.g. with *main = "..."* for the title.

The histogram shows how often individual values occur in certain classes. In addition, we often use the histogram to check whether a variable is normally distributed (see Chap. 5). We display the histogram with the hist() command. For example, we want to analyze the distribution of guests according to their age. The command is:

```
> hist(tourism$age)
```

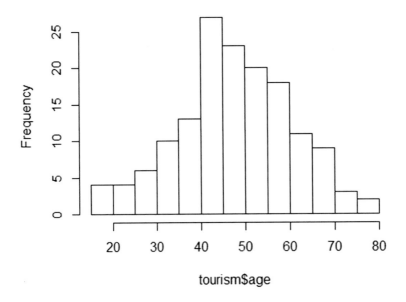

Histogram of tourism$age

We see how many guests are present in which age category and that the age category from 40 to 45 is recorded most frequently.

If we would like to add a title, label the axes and perhaps even change the color of the bars, we add *main = "..."* for the title of the graph, *xlab = "..."* and *ylab = "..."* for the axis titles and *col = "..."* for the color of the bars. The command could then look like this.

```
> hist(tourism$age, main = "Guests in age categories", xlab = "Age", ylab =
"Frequency", col = "orange")
```

Guests in age categories

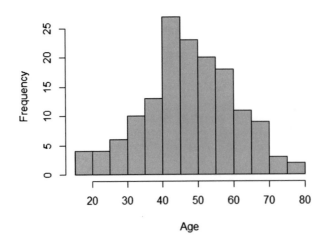

Another frequently used graphic is the pie chart. Usually we draw a pie chart when we show how often individual values of a nominal or ordinal variable occur. The command for a pie chart is *pie()*. As an example, we would like to show from which countries our guests come. First we count the frequencies with the function *table()*. Then we integrate the function into the command *pie()*.

```
> table(tourism$country)

##
##  1  2  3  4
## 64 48 26 12
```

```
> pie(table(tourism$country))
```

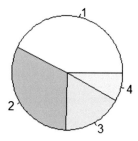

We see that most of the guests come from the country with the coding 1, followed by the countries with the coding 2, 3 and 4. If we now still know our coding (legend of the dataset: 1 = Switzerland; 2 = Germany; 3 = Austria; 4 = others) then we see that most of the guests come from Switzerland, followed by Germany and Austria.

If we want to display the names of the countries of origin in the diagram instead of the coding, this will be done with the addition *labels = c()*. Note that each label name is separated by a comma and is enclosed in quotation marks, and that the ordering matches the coded digits. The command now looks like this:

```
> table(tourism$country)

##
##  1  2  3  4
## 64 48 26 12
```

```
> pie(table(tourism$country), main = "Origin of guests", labels =
c("Switzerland", "Germany", "Austria", "others"))
```

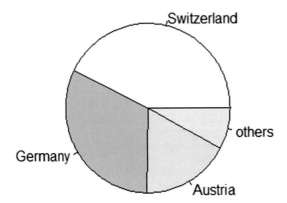

Another important diagram is the bar chart. We usually use the bar chart as an alternative to the pie chart or when we compare groups. To obtain a bar chart, we use the same procedure as for the pie chart. We first count the numbers with the *table()* function and then integrate the function into the *barplot()* function. The columns are labeled with the extension *names.arg = c()*. The command is as follows:

```
> table(tourism$education)

##
##  1  2  3  4
## 59 32 32 27
```

```
> barplot(table(tourism$education), main = "Education level of guests", xlab
= "Highest degree", ylab = "Frequency", names.arg = c("Sec", "A-
Level","Bachelor","Master"), col = "orange")
```

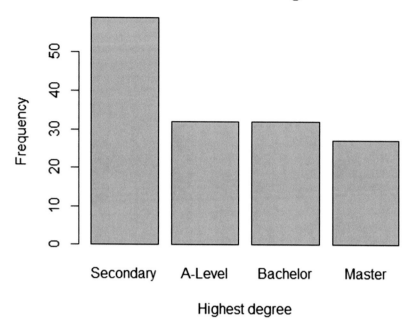

We see in the command that we have labeled the axes, given the bars a color and named the columns.

Bar charts are particularly suitable for group comparisons. We are still interested in the education level, but want to analyze it separately for women and men.

First we count the frequencies separately by gender with the function *table()*. Then we integrate the function into our *barplot()* command. We can do some things in addition. If we want the columns to be next to each other, we have to add *beside = TRUE*. Furthermore, it makes sense to assign colors to the columns. We accomplish this with *col = c* ("...", "..."). A legend is also necessary so that we know which bar belongs to which group. The extension legend = ... is used. The command is then as follows:

```
> table(tourism$sex, tourism$education)

##
##      1  2  3  4
##   0 31 20 18 16
##   1 28 12 14 11


> barplot(table(tourism$sex, tourism$education), main = "Education level of
guests", xlab = "Highest degree", ylab = "Frequency", names.arg = c("Sec", "A-
Level", "Bachelor", "Master"), beside = TRUE, col=c("darkblue","red"), legend
= rownames(table(tourism$sex, tourism$education)))
```

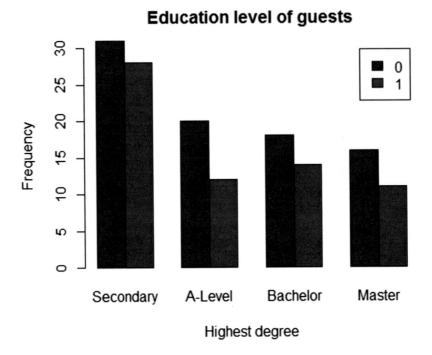

Finally, we want to have a look at the boxplot and the scatterplot. The boxplot is used to evaluate the variation of variables and groups and to identify outliers. The scatterplot is used to analyze the relationship between two variables and to identify unusual observations.

To get the boxplot, we enter the command *boxplot()*. Here we specify the variable for which we want to create the boxplot and, if necessary, the grouping variable for which we want to create the boxplot. The grouping variable is separated by a tilde character ∼:

```
> boxplot(tourism$age ~ tourism$sex,  main = "Variation of the variable age by
sex", xlab = "Sex", ylab = "Age", names=c("male", "female"))
```

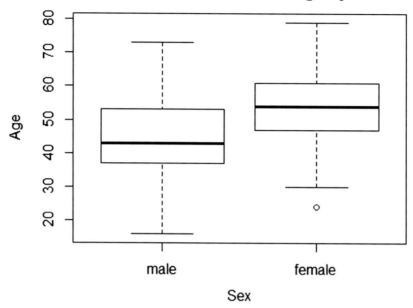

The simple scatterplot between two variables is obtained with the *plot()* function and the respective variable names:

```
> plot(tourism$satisfaction, tourism$expenses, main = "Relationship between
expenses and satisfaction", xlab = "Satisfaction with the ski destination",
ylab = "Expenses in CHF per day")
```

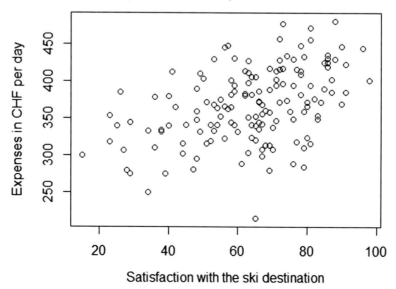

If we also want to have a line in the diagram fitting to the data cloud, we use the *scatter.smooth()* function.

```
> scatter.smooth(tourism$satisfaction, tourism$expenses, main = "Relationship
between expenses and satisfaction", xlab = "Satisfaction with the ski
destination", ylab = "Expenses in CHF per day")
```

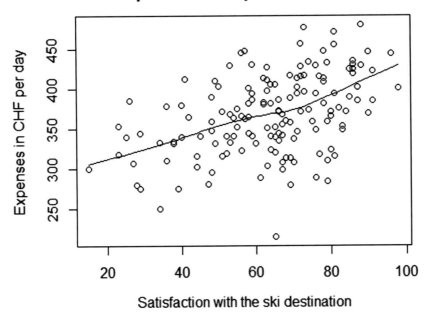

Statistical Charts with ggplot2

If we have the task to create presentation-ready or more complex charts, then it makes sense to use an appropriate package. A well-known package for graphs is *ggplot2*. To use it, we have to activate it; if necessary, we have to install it first.

```
> library(ggplot2)
```

At this point, the logic of *ggplot2* has to be described briefly. The package assumes that a graphic can be represented by several layers. The function *ggplot()* starts the basic layer, which can be completed with a "+" by further layers. Data points are mapped with the function *geom_*. In the following, we want to show this principle with the already introduced graphics. At this point, we have to point out that the script is only a simple introduction and by no means discusses all functions of *ggplot2*.

To draw the histogram with *ggplot2*, we first open the basic layer for the required graphic with the command *ggplot(tourism, aes(age))*. Within the command we specify which dataset should be used and the variable with the extension *aes(age)*. Then we extend the function with *geom_histogram()*. We tell RStudio that a histogram should be drawn into the base level. If we leave the parenthesis empty, we get the standard version. The histogram is gray with an automatically generated number of classes. In the following we add *binwidth = 5*, *fill = "white"*, *color = "blue"*, i.e. the width of the classes is five units, the columns are filled in white and bordered in blue. Additionally we define the title and the axis label with + *ggtitle("Histogram for the variable age") + xlab("Age") + ylab("Frequency")*.

```
> ggplot(tourism, aes(age)) + geom_histogram(binwidth = 5, fill = "white",
colour = "blue") + ggtitle("Histogram for the variable age") + xlab("Age") +
ylab("Frequency")
```

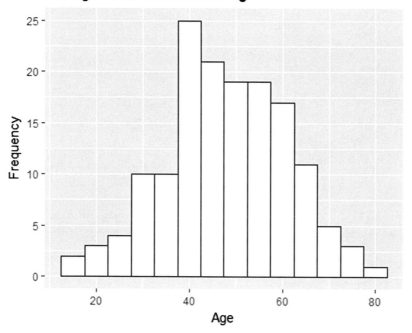

The pie chart is a little bit more complicated to implement in *ggplot2*, so instead we show how the bar chart is generated. The bar chart can be used to display the same information as the pie chart.

To create a bar chart, we open the base level again with *ggplot()* and then extend the function with *geom_bar()*. The variable *country* has to be recoded into a factor first.

```
> f_country <- as.factor(tourism$country)
> tourism <- cbind(tourism, f_country)

> ggplot(tourism, aes(f_country)) + geom_bar(fill = "orange") +  ggtitle("Bar
chart for the variable country") + xlab("Country") + ylab("Frequency") +
scale_x_discrete(breaks = c("1","2","3","4"), labels =
c("CH","GER","AUT","others"))
```

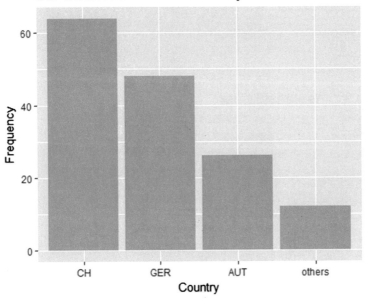

If we would like to analyze the education level separately by gender with the help of the bar chart, we proceed in the same way. As always, the command adapts slightly.

```
> f_education <- as.factor(tourism$education)
> f_sex <- as.factor(tourism$sex)
> tourism <- cbind(tourism, f_education, f_sex)

> ggplot(tourism, aes(f_education)) + geom_bar(aes(fill = f_sex),
position="dodge") + ggtitle("Bar chart for education by gender") +
xlab("Education level") + ylab("Frequency") + scale_x_discrete(breaks =
c("1","2","3","4"), labels = c("Secondary","A-Level","Bacherlor","Master"))
```

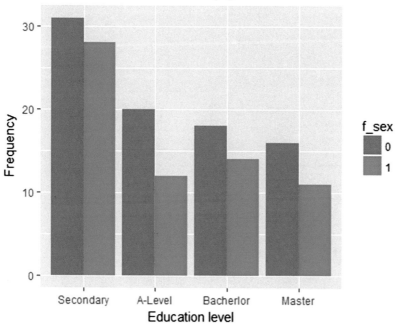

We get the boxplot with the addition *geom_boxplot()*. It is important that we specify the dataset, the grouping variable and the variable of interest within the *ggplot()* function. We already know the rest. The function is shown below:

```
> ggplot(tourism, aes(f_sex, age)) + geom_boxplot(outlier.shape = 8,
outlier.colour = "red") + ggtitle("Boxplot for age by gender") +
xlab("Gender") + ylab("Age in years") + scale_x_discrete(breaks = c("0",
"1"), labels = c("male","female"))
```

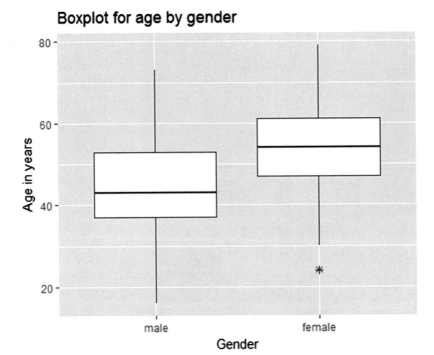

Finally we show the scatterplot with *ggplot()*. Let us assume that there is a relationship between the satisfaction of the guests and their expenses and that we want to display this graphically. The command is as follows, we specify both variables and use +*geom_point()* as extension:

```
> ggplot(tourism, aes(x=expenses,y=satisfaction))+geom_point() + ggtitle
("Scat terplot for satisfaction and expenses") + xlab("Expenses in CHF
per day")+ ylab("Satisfaction with the ski destination")
```

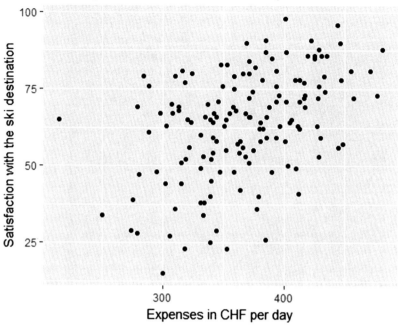

There seems to be a slightly positive correlation between the variables *expenses* and *satisfaction*. We now wish to make the graph more appealing. For example, we want to make the dots blue and a little bit larger. With *color= "..."* we define the desired color, with *alpha* its opacity and with size the *size* of the points.

```
> ggplot(tourism, aes(x=expenses,y=satisfaction)) + geom_point(colour="blue",
alpha=0.5, size=2.0) + ggtitle("Scatterplot for satisfaction and expenses") +
xlab("Expenses in CHF per day") + ylab("Satisfaction with the ski
destination")
```

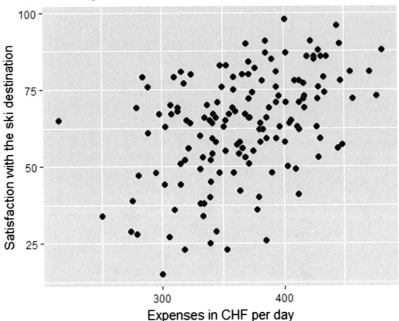

The graph is now a little bit more colorful. Now we would like to examine whether women and men differ in terms of expenses and satisfaction. For this, the symbols in the scatterplot should differ by gender and each gender should have a different color. The command to use is similar to the command above, but we add directly to the axis information that *color* and *shape* should be displayed differently according to gender.

```
> f_sex <- as.factor(tourism$sex)
> tourism <- cbind(tourism, f_sex)

> ggplot(tourism, aes(x=expenses,y=satisfaction, colour=f_sex,
shape=f_sex))+geom_point(size=1.5) + ggtitle("Scatterplot for satisfaction
and expenses") + xlab("Expenses in CHF per day") + ylab("Satisfaction with
the ski destination")
```

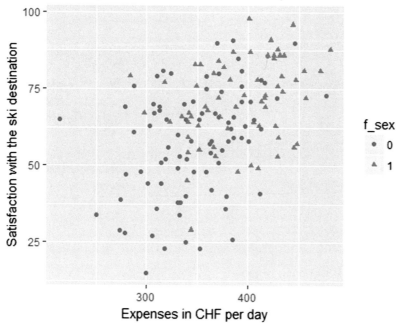

To see the graphic in full size, we click on *Zoom* in the graphics window in the upper menu bar.

The interested reader is invited to extend his or her abilities to create graphics with R by himself or herself. A further description would make this easy introduction too complex.

But we would like to show how the produced graphs are saved. To do this, we click on *Export* in the upper menu bar in the graphics window. We can choose whether the graph should be saved as a PDF or as an image file (selection of different formats) and specify the desired target folder (Fig. 4.1: Export graphs).

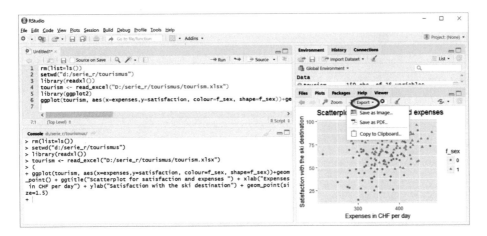

Fig. 4.1 Export graphs

Table 4.1 R commands learned in this chapter

Command	Notes/examples
barplot()	....
boxplot()	....
cor()	....
ggplot()	....
hist()	....
mean()	....
median()	....
numSummary()	....
pie()	....
plot()	....
quantile()	....
scatter.smooth()	....

(continued)

Table 4.1 (continued)

Command	Notes/examples
sd()	....
	....
summary()	....
	....
table()	....
	....
var()	....
	....

Note You can take notes or write down examples for their usage on the right side

4.3   Time to Try

Use for the following tasks the dataset tourism.xlsx.

4.1. Find the modus for the variables *accommodation*, *education* and *stay* and interpret the results. Do not forget to use the legend of the data, when interpreting the results.

4.2. Display the median and the mean for the variables *age*, *expenses* and *stay*. Interpret the results.

4.3. Calculate the standard deviation, the variance and the quantiles for the variables *age*, *expenses* and *stay*. Interpret the results together with the results of task 2.

4.4. Display the median, mean, standard deviation, variance, inter quartile range, the smallest value, largest value and the first and the third quartile for the variables *diversity*, *quality*, *safety*, *satisfaction* and *waitingtime* with one command and compare the results.

4.5. Figure out how women and men feel about the *waitingtime* at the ski lifts. Is there a difference between them in the sample regarding the average feeling and the variation?

4.6. Calculate the correlation coefficient of Bravais-Pearson between the two variables *age* and *expenses*. Interpret the result.

4.7. Calculate the correlation coefficients of Bravais-Pearson between the variables *age*, *expenses* and *stay* using the correlation matrix. Interpret the results.

4.8. Calculate the correlation coefficient of Spearman between the two variables *accommodation* and *education*. Interpret the result.

4.9. Calculate the correlation coefficients of Spearman between the variables *accommodation*, *recommendation* and *education* using the correlation matrix. Interpret the results.

4.10. Create a histogram for the variable *expenses* and interpret the graph briefly. Try to label the histogram, the x-axis and the y-axis and give the histogram the color purple.

4.11. Create a pie chart for the variable education and interpret the pie chart briefly. Try to add the names to the pieces.

4.12. Try to create a barplot that displays the guests by country and accommodation category. Label the barplot properly and interpret the result.

4.13. Create a boxplot that shows the expenses by the country of the guests. Label the boxplot and interpret the results.

4.14. Draw a scatterplot for the variables *age* and *expenses*. Label it and interpret the result.

4.15. Try to draw a smooth line into the graph of task 14. Is there a linear relationship (straight line) between the two variables?

Testing Normal Distribution with RStudio

5

In many statistical applications, the question arises whether the variable of interest is normally distributed. Normal distribution is often a prerequisite for the application of statistical methods.

To check whether a variable is normally distributed, RStudio provides a number of methods that can be used. We'll show you four options: with the histogram and the quantile quantile plot (Q-Q plot) two graphical ones and with the skewness and the kurtosis and the Shapiro-Wilk test two numerical ways.

Before we start, we delete our working memory again and load the required data:

```
> rm(list=ls())
> library(readxl())
> tourism <- read_excel("D:/serie_r/tourismus/tourism.xlsx")
```

5.1 Graphical Ways to Check for Normal Distribution

The histogram shows how many values we have per class. A variable is considered to be normally distributed if the histogram is symmetrical, not too flat, and not too peaked. Simply imagine a church bell (Gaussian bell curve) fitting into the histogram. As shown above, we draw the histogram itself with the *hist()* command.

```
> hist(tourism$age)
```

Histogram of tourism$age

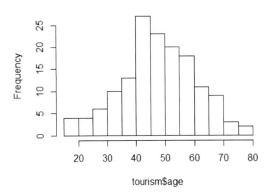

tourism$age

Let's have a quick look at the histogram. The variable doesn't seem to be 100% symmetrical, the columns are slightly different in height and if the histogram is folded in the middle, the two sides won't lie on top of each other. An imaginary bell does not fit into the histogram. Hence, we have a first hint.

The quantile-quantile plot compares the theoretical quantiles of a normally distributed variable (with the mean value and the standard deviation of the variable to be checked) with the actual observed quantiles of the variable to be examined. The command to create the diagram is as follows, if the package *car* is active:

```
> library(car)
```

```
> qqPlot(tourism$age)
```

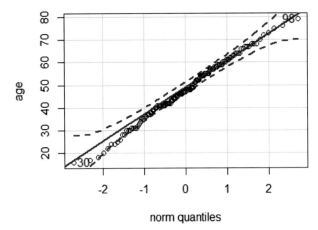

norm quantiles

```
## [1] 30 98
```

Let's take a quick look at the diagram. If the theoretically calculated quantiles correspond to the empirically observed quantiles, then the points should all be on the line. If this is not the case, the variable deviates from the normal distribution. With the command we get not only the line, but also a confidence range (dashed lines), i.e. if the points lie within the dashed lines, we assume that the variable is normally distributed.

Here are some reasons a variable might deviate from the normal distribution line. This may be helpful if we want to transform the variable to achieve a normal distribution (see chapter Regression Diagnostics).

Pattern	Interpretation
All points are on the line, besides a few exceptions	There are outliers in the data
The left end of the points is below the line, the right end is above the line	The distribution is peaked
The left end of the points is above the line, the right end is below the line	The distribution is flat
The slope of the curve increases from left to right	The distribution is right skewed
The slope of the curve decreases from left to right	The distribution is left skewed

5.2   Numerical Ways to Check for Normal Distribution

Besides the graphical evaluation, we can also carry out numerical investigations. A normally distributed variable is symmetric with a skewness of 0 and not too peaked (leptokurtic) and too flat (platykurtic) but bell-shaped with an kurtosis of 3 and an excess kurtosis of 0. Note we receive the excess kurtosis when we subtract 3 from the kurtosis. Hence, when we calculate the skewness and the excess kurtosis we should find values close to 0.

To receive the skewness and the excess kurtosis we need the package e1071 and the commands *kurtosis()* and *skewness()*. It is often a little bit confusing that with the command kurtosis we will obtain in fact the excess kurtosis. If you are unsure about it then best have a look at the histogram to cross check. The commands are as follows:

```
> library(e1071)
```

```
> kurtosis(tourism$age)
```

```
## [1] -0.2994913
```

```
> skewness(tourism$age)
```

```
## [1] -0.1085657
```

In this case, the excess kurtosis as well as the skewness are close to 0, i.e. the variable is not far away from a normal distribution.

Finally, the Shapiro-Wilk test checks whether the variable is normally distributed in its population. The null hypothesis states that the variable is normally distributed in the population. The alternative hypothesis is that the variable is not normally distributed in the population. The test is usually performed at a significance level of 5%, i.e. if we have a p-value of less than 5%, we reject the null hypothesis.

The command to call the Shapiro-Wilk test is as follows:

```
> shapiro.test(tourism$age)

##
##   Shapiro-Wilk normality test
##
## data:  age
## W = 0.99325, p-value = 0.7075
```

In this case, the p-value is clearly greater than 5% (0.7075 > 0.05), we do not reject the null hypothesis; we assume that the variable is normally distributed in its population.

We have now learned four ways to check whether a variable is normally distributed. Each of these methods alone is not sufficient for an assessment. It is best to perform all four procedures and finally assess the results of the procedures as a whole. Based on this evaluation we make the decision. In our case, we would conclude that the variable is normally distributed. The Shapiro-Wilk test and the quantile-quantile plot indicate this result, the histogram, the skewness and the excess kurtosis deviate only slightly from it.

Table 5.1 R commands learned in this chapter

Command	Notes/examples
histogram()	....
kurtosis()	....
qqPlot()	....
shapiro.test()	....
skewness()	....

Note You can take notes or write down examples for their usage on the right side

5.3 Time to Try

5.1. Check for the variable *age* of the dataset dogs.xlsx whether the variable follows a normal distribution.
5.2. Check for the variable *expenses* of the dataset tourism.xlsx whether the variable is normally distributed.
5.3. Check for the variable *satisfaction* of the dataset tourism.xlsx whether the variable is normally distributed.

Testing Hypotheses with RStudio

6

The following chapter shows how to use RStudio to test hypotheses. Hereby, the decision tree (Fig. 6.1: Guide to select tests) will help to decide which test should be used. We see that it is crucial to know whether we test for group differences or relationships between

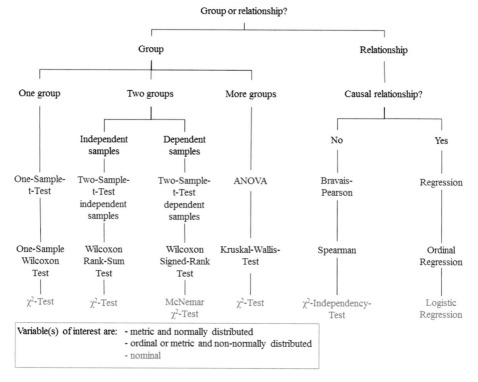

Fig. 6.1 Guide to select tests

© The Author(s), under exclusive license to Springer-Verlag GmbH, DE, part of Springer Nature 2021
F. Kronthaler and S. Zöllner, *Data Analysis with RStudio*,
https://doi.org/10.1007/978-3-662-62518-7_6

variables and whether the test variable is metric, ordinal or nominal. Furthermore, the question arises whether metric test variables are normally distributed.

In the following, we will not introduce all test techniques, but will only deal with frequently used tests. The script is not intended to discuss all test techniques, but to provide an easy introduction to working with RStudio. At this point we would like to remind you that additional literature should be consulted for statistical questions and further questions on R or RStudio (Chap. 9: Further Reading).

We will now start and load our data:

```
> rm(list=ls())
> library(readxl())
> tourism <- read_excel("D:/serie_r/tourismus/tourism.xlsx")
```

6.1   One-Sample t-Test

A frequently used test method is the one-sample t-test. With the one-sample t-test, we test whether an assumption (hypothesis) about the average behavior of a group is justified. For example, we could assume that girls at the age of 16 smoke on average 5 cigarettes per day. With the help of a sample and the one-sample t-test, we can examine this assumption.

For example, if we take the *tourism* dataset available, we could check how much money guests spend on average per day. Let's assume that we have a guess, we know, for example, that five years ago guests spent CHF 300 per day on average (including overnight stays). We want to check whether this amount has changed since then. The null hypothesis in this case would be that the average expenditure per guest and day is CHF 300. The corresponding alternative hypothesis is that the average expenditure per guest and day is higher or lower than CHF 300. We test two-sided because we are "only" interested in a change, which can be negative or positive. Before we test, we should determine the level of significance. Let's assume that we choose 5%. The command for the test is as follows:

```
> t.test(tourism$expenses, alternative = "two.sided", mu = 300, conf.level =
0.95)

##
##  One Sample t-test
##
## data:  tourism$expenses
## t = 16.663, df = 149, p-value < 2.2e-16
## alternative hypothesis: true mean is not equal to 300
## 95 percent confidence interval:
##   359.8129 375.9071
## sample estimates:
## mean of x
##    367.86
```

Let's take a quick look at the command and the result. The *t.test()* function specifies that we perform a t-test for a metric and normally distributed variable. Next we specify the variable *tourism$expenses* and tell RStudio that we want the two-sided test *alternative = c ("two.sided")*. With *mu = 300* we tell RStudio that the mean value which we test is CHF 300. Finally, we specify with *conf.level = 0.95* that we want to see the 95% confidence interval in addition to the test.

In the result output, we see the calculated t-value of the test statistics, the number of degrees of freedom and the p-value. If the p-value is less than the specified significance level, we reject the null hypothesis and proceed with the alternative hypothesis. We also see the 95% confidence interval, which ranges from CHF 360 to CHF 376, and the mean value of the sample data of CHF 368.

6.2   Two-Sample t-Test Independent Samples

With the t-test for independent samples, we test whether two independent groups behave differently. For example, we test whether girls aged 16 on average smoke more or less cigarettes than boys aged 16.

With the help of the tourism dataset, we could test, for example, whether women are more satisfied with our ski resort than men are. The null hypothesis is that on average women are less satisfied or equally satisfied with the ski resort than men are. The alternative hypothesis is that on average women are more satisfied with the ski resort than men are. We want to test at the significance level of 10%. The command is as follows:

```
> f_sex <- as.factor(tourism$sex)
> tourism <- cbind(tourism, f_sex)

> t.test(satisfaction ~ f_sex, data = tourism, alternative = "less", mu = 0,
paired = FALSE, var.equal = TRUE, conf.level = 0.95)

##
##   Two Sample t-test
##
## data:  tourism$satisfaction by f_sex
## t = -4.7191, df = 148, p-value = 2.719e-06
## alternative hypothesis: true difference in means is less than 0
## 95 percent confidence interval:
##        -Inf -8.133008
## sample estimates:
## mean in group 0 mean in group 1
##        58.21176        70.73846
```

Before we are able to test, we have to define gender as a factor. Furthermore, we have to think about how men and women are coded in the dataset. This is necessary because we test one-sided and we have to decide whether we have to test left- or right-sided. In the data set, the women are coded with a 1 and the men with a 0 (legend of the dataset).

If we look at the *t.test()* function, we see that we first specify the test variable *tourism$satisfaction*. Then we tell RStudio that the group variable is the factor ~*f_sex*. With the next addition we tell RStudio that we test left-sided, with the difference *mu = 0*. With *paired = FALSE* we specify that we have independent groups and with *var. equal = TRUE* we tell RStudio that the variances in both groups are the same (we did not test this here). Finally, with *conf.level = 0.95* we ask for the 95% confidence interval.

In the output, we first see the t-value of the test statistics, then the number of degrees of freedom and then the p-value. The p-value is less than 10%, i.e. we reject the null hypothesis. Women are more satisfied with the ski resort than men (for whatever reason). The 95% confidence interval ranges from minus infinity to −8.1. The sample value for men is roughly 58 points and the sample value for women 71 points.

6.3 Wilcoxon Rank-Sum Test

The Wilcoxon rank-sum test is the alternative to the two-sample t-test for independent samples when the test variable is metric but not normally distributed or ordinal. If we assume that the satisfaction variable is not normally distributed, the Wilcoxon rank sum test would come into use instead of the t test for independent samples (Fig. 6.1: Guide to select tests). The command for the same test situation used before is now:

```
> f_sex <- as.factor(tourism$sex)
> tourism <- cbind(tourism, f_sex)

> tapply(tourism$satisfaction, tourism$f_sex, median, na.rm=TRUE)

## 0  1
## 61 72
```

```
> wilcox.test(satisfaction ~ f_sex, data = tourism, alternative="less")

##
##   Wilcoxon rank sum test with continuity correction
##
## data:  tourismus$satisfaction by f_sex
## W = 1640.5, p-value = 1.048e-05
## alternative hypothesis: true location shift is less than 0
```

With the function *tapply()* RStudio is instructed to table values for the variable *satisfaction*. *f_sex* specifies that we would like to have the values for man and women. *median* specifies that we would like to have the median for both groups. With *na.rm=TRUE* we tell RStudio how to handle missing values (more in Appendix C).

With *wilcox.test()* the Wilcoxon rank-sum test is started. *tourism$satisfaction* is the dataset and the test variable and the grouping variable is *f_sex*. Finally we specify the direction of the hypothesis test with *alternative="less"*.

Let's take a look at the results. First, we notice that the median for group 0, the men, is with 61 smaller than for group 1, the women, with 72. Then we get the W, which is our test statistic, as well as the p-value. The p-value is our criterion for whether we reject the null hypothesis. In this case, the p-value is substantially lower than our defined 10% significance level, i.e. we reject the null hypothesis and conclude that women are more satisfied with the ski resort than men are.

6.4   Two-Sample t-Test Dependent Samples

Now let's briefly discuss the t-test for dependent samples. It tests whether an event or a measure has an influence on objects or persons. For this to be tested, the objects or persons must be observed twice, once before a treatment and once after the treatment. Our data set used so far is not sufficient for this, as we interviewed the persons only at one point of time. Therefore, we need another dataset. The dataset *schoolbreak* contains a suitable example. In this fictitious dataset, schoolchildren are observed twice, once before the school has introduced an active school break with sports units and once a month later after the school has introduced active school breaks. The question in this example is quite simple, whether active school breaks help to reduce the (over-)weight of schoolchildren. The null hypothesis is that the weight has not changed with the measure or has become higher. The alternative hypothesis is that the weight has decreased after the measure. We want to test at a significance level of 5%. Before we can perform the t-test for dependent samples, we must load the dataset.

```
> library(readxl)
> schoolbreak <- read_excel("D:/serie_r/tourismus/schoolbreak.xlsx")
```

After loading the dataset, we can perform the t-test for dependent samples. The command is as follows:

```
> t.test(schoolbreak$before, schoolbreak$after, alternative = "greater", mu =
0, paired = TRUE, conf.level = 0.95)

##
##   Paired t-test
##
## data:   schoolbreak$before and schoolbreak$after
## t = 2.3702, df = 8, p-value = 0.02262
## alternative hypothesis: true difference in means is greater than 0
## 95 percent confidence interval:
##  0.3662432        Inf
## sample estimates:
## mean of the differences
##                      1.7
```

Let's take a quick look at the command. The function is the *t.test()* function again. We have specified two variables *schoolbreak$before* and *schoolbreak$after* (legend of the dataset). In this case we test the right-sided *alternative = c("greater")*. The assumed difference is *mu =0* and with *paired = TRUE* we tell RStudio that we have dependent samples. Furthermore, with *conf.level = 0.95* we request the 95% confidence interval.

Let's take a look at the result. First, we see the t-value of the test statistic, then the number of degrees of freedom and then the p-value. The p-value is less than 5%, i.e. we reject the null hypothesis. The measure helps to reduce the weight. The 95% confidence interval ranges from 0.37 to infinity. The difference in the sample is 1.7, i.e. the mean value in the sample was 1.7 kg higher before the measure than after the measure.

6.5 Wilcoxon Signed-Rank Test

The Wilcoxon signed-rank test is used when we have dependent samples and the test variable is either metric but not normally distributed or ordinal. Let's take the example we just used with the school break. Let us now assume that the test variable is metric, but not normally distributed. In this case, we take the Wilcoxon signed-rank test instead of the t-test for dependent samples. Everything else stays the same; the command is then as follows:

```
> median(schoolbreak$before)

## [1] 32.5

> median(schoolbreak$after)

## [1] 30.9

> wilcox.test(schoolbreak$before, schoolbreak$after, alternative='greater',
paired=TRUE)

##
##  Wilcoxon signed rank test
##
## data:  schoolbreak$before and schoolbreak$after
## V = 39, p-value = 0.02734
## alternative hypothesis: true location shift is greater than 0
```

Median() is used to display the medians for the two points in time. With *wilcox.test()* we start the Wilcoxon signed-rank test. Then we give RStudio the variables *schoolbreak$before* and *schoolbreak$after*. We specify the direction of the hypothesis test with *alternative* = *"greater"*, furthermore we tell R with *paired* = *TRUE* that it is a paired sample.

Let's have a look at the result. First, we see that the median was higher in the sample before the introduction of the active school break than after the introduction of the active school break. In the test output, we get two variables, the test statistic V and the p-value. The p-value is less than 5%, from which we conclude at the significance level of 5% that we reject the null hypothesis. The active school break thus helps to reduce the weight.

6.6 Analysis of Variance ANOVA

So far, we have only compared two groups. But what do we do if we want to test more than two independent groups at the same time? In such a case, the analysis of variance is the method to be used. The analysis of variance is applied when we have more than two groups for one grouping variable or more than one grouping variable at the same time.

At this point, we would briefly like to refer to the assumptions of analysis of variance without discussing them in more detail. The assumptions are (1), the test variable is metric and normally distributed, (2) the groups are independent from each other, and (3) the groups have equal variances.

Here comes a small example for illustration. Let's assume we're analyzing beer consumption at the Oktoberfest. We are interested in whether Japanese, Bavarians or Italians drink more beer per person and whether gender also has an influence. The test variable is beer consumption and the grouping variables are country of origin and gender. In the analysis of variance, the grouping variables are also called factors or independent variables. The test variable is also called dependent variable. The assumption (1) requires that beer consumption was measured metrically and that the variable is normally distributed, the assumption (2) requires that the groups drink independently of each other and the assumption (3) assumes equal variation for the groups.

We now want to go through an example with the help of our dataset. We are interested in whether there is a difference in the daily expenses between Swiss, Germans, Austrians or guests from other countries. Therefore, we have a metric test variable and one factor, the variable *country*, which specifies which country the guests come from.

The null hypothesis in the analysis of variance is that there are no group differences; the alternative hypothesis is accordingly, there are group differences. We want to test at the 5% significance level.

Before we start with the analysis of variance, we should test the assumptions (we omitted this step in the test techniques above). As shown in Sect. 6.5, we have several options to test for normal distribution, such as the histogram, the Q-Q plot, and the Shapiro-Wilk test. The commands are as follows:

```
> hist(tourism$expenses)
```

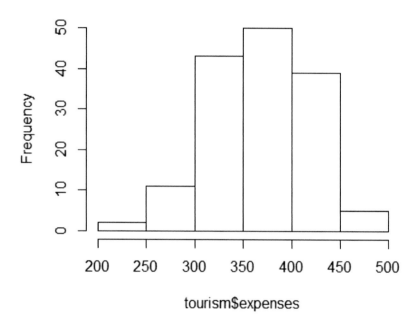

Histogram of tourism$expenses

```
> library(car)
```

```
> qqPlot(tourism$expenses)
```

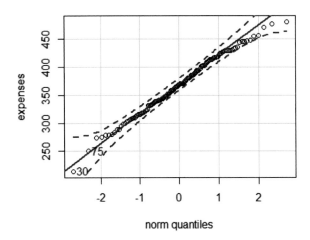

```
## [1] 30 98
```

```
> shapiro.test(tourism$expenses)

##
##  Shapiro-Wilk normality test
##
## data:  expenses
## W = 0.9932, p-value = 0.7026
```

The results indicate that the variable *expenses* is normally distributed in its population. So everything is fine with the normal distribution assumption. If we have a problem with it, we could look at our figure in Sect. 6.6 to see if there is an alternative to the analysis of variance. We would see that the Kruskal-Wallis test is the appropriate alternative for a metric but not-normally distributed test variable.

The assumption of independence of groups cannot be tested. This is a question to be answered theoretically or during data collection. What remains is the assumption of the same group variances. Here, too, we have various alternatives. Popular options are the boxplot and the Levene test for equal variances. The boxplot shows us the variation of the groups graphically, the Levene test tests for equal group variances with the null hypothesis, the group variances are equal, and the alternative hypothesis, the group variances are not equal. Usually a significance level of 5% is chosen.

The commands are as follows when the *RcmdrMisc* package is active:

```
> library(RcmdrMisc)
> f_country <- as.factor(tourism$country)
> tourism <- cbind(tourism, f_country)
> Boxplot(expenses~f_country, data=tourism)
```

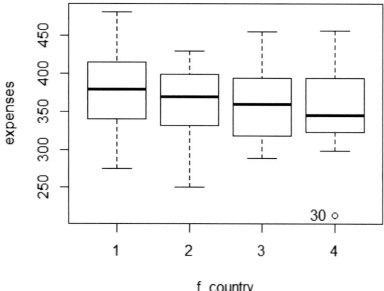

```
## [1] "30"
```

```
> with(tourism, tapply(expenses, f_country, var, na.rm=TRUE))
```

```
##        1        2        3        4
## 2467.286 2049.851 2390.375 4427.841
```

```
> leveneTest(expenses ~ f_country, data=tourism, center="mean")
```

```
## Levene's Test for Homogeneity of Variance (center = "mean")
##          Df F value Pr(>F)
## group     3  0.3876 0.7621
##         146
```

Let's take a look at the commands. First, of course, we have to tell RStudio that the grouping variable *country* is a factor. Then we can use the *Boxplot()* command to display the boxplots for the groups. With the function *tapply()* we get the variances for the groups and with the *leveneTest()* we select the test for equal variances.

The variances are of course slightly different. If we look at the boxes of the groups, however, it becomes clear that the middle 50% of the values scatter in approximately the same bandwidth. Furthermore, the Levene test is not significant, i.e. we do not reject the null hypothesis and assume equal variances.

The assumptions are now tested and we can perform the analysis of variance. The commands are:

```
> AnovaModel.1 <- aov(expenses ~ f_country, data=tourism)
> summary(AnovaModel.1)

##             Df Sum Sq Mean Sq F value Pr(>F)
## f_country    3  10412    3471   1.407  0.243
## Residuals  146 360248    2467

> TukeyHSD(AnovaModel.1)
##    Tukey multiple comparisons of means
##      95% family-wise confidence level
##
## Fit: aov(formula = expenses ~ f_country, data = tourism)
##
## $f_country
##              diff       lwr      upr     p adj
## 2-1 -14.87500000 -39.52441  9.774409 0.3999212
## 3-1 -14.97115385 -44.99406 15.051750 0.5669846
## 4-1 -23.87500000 -64.48512 16.735122 0.4235293
## 3-2  -0.09615385 -31.53139 31.339078 0.9999998
## 4-2  -9.00000000 -50.66511 32.665106 0.9432930
## 4-3  -8.90384615 -53.95673 36.149039 0.9557302
```

```
> with(tourism, plotMeans(expenses, f_country, error.bars="conf.int",
level=0.95))
```

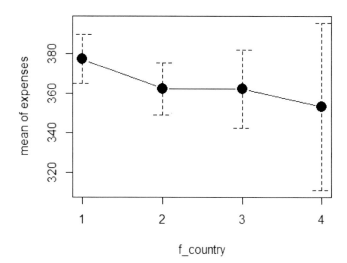

Plot of Means

The first command tells RStudio to perform an analysis of variance. The assignment operator <- specifies that the results should be stored in the *AnovaModel.1* object. After the assignment operator, *aov()* "analysis of variance" is the actual command for the variance analysis. First the variable of interest is specified in the command (*expenses*), then the grouping variable/factor is listed separated by the tilde character ∼. The last part of the command specifies the data set to which the command applies.

In the next line, we instruct RStudio with *summary()* to display the results of the analysis of variance. The *TukeyHSD()* command in the third line tells RStudio to perform a post-hoc test. This has the following background: the analysis of variance tests for group differences. The analysis of variance does not test which groups are different. In order to determine the latter, it is necessary to analyze which groups are different. This is done with so-called post-hoc tests. A popular post-hoc test is the Tukey test. With the last command *plotMeans* we display the results graphically.

Let us now look at the output block of the analysis of variance. We get several values for our grouping variable/factor *f_country* and the so-called *residuals*. The values are the degrees of freedom (Df), the sum of squares (Sum Sq), the mean sum of squares (Mean Sq), the test statistic (F value) and the p-value (Pr(>F)).

Let us concentrate on the sum of squares (Sum Sq) and on the p-value (Pr(>F)). In the column *Sum Sq* we see the sum of squares explained by the factor *f_country* and the sum of squares not explained by the factor (residuals). Altogether, only a small part of the total measured variance (10412+360248) is explained. Accordingly, the p-value is 23.4% relatively high, and we do not reject the null hypothesis at the 5% significance level. This means that there are no group differences and that Swiss, Germans, Austrians and guests from other nations spend the same amount of money. Since there are no group differences overall, we do not need to worry about the post-hoc test anymore. But here, too, if we look at the p-value (p adj), we see that none of the group comparisons shows a value below the 5% significance level. Graphically we also see this of course in the mean plot, the mean values are relatively similar and the error bars overlap.

As already mentioned above, the analysis of variance can be extended easily by further factors. As an example, we want to perform an analysis of variance with two factors *f_country* and *f_sex*. The command only has to be extended by the second variable/factor.

Before we perform the analysis of variance, we first delete the working memory. There is currently no specific reason for this, except that it reminds us of the command.

```
> rm(list=ls())
```

Then, of course, we have to reload the data before we can execute the commands:

```
> library(readxl)
> tourism <- read_excel("D:/serie_r/tourismus/tourism.xlsx")
> f_country <- as.factor(tourism$country)
> f_sex <- as.factor(tourism$sex)
> tourism <- cbind(tourism, f_country, f_sex)
> AnovaModel.2 <- aov(expenses ~ f_country*f_sex, data=tourism)
> summary(AnovaModel.2)

##                   Df Sum Sq Mean Sq F value    Pr(>F)
## f_country          3  10412    3471   1.687     0.173
## f_sex              1  64345   64345  31.272  1.11e-07 ***
## f_country:f_sex    3   3719    1240   0.602     0.614
## Residuals        142 292184    2058
## ---
## Signif. codes:  0 '***' 0.001 '**' 0.01 '*' 0.05 '.' 0.1 ' ' 1
```

```
> TukeyHSD(AnovaModel.2)

##    Tukey multiple comparisons of means
##      95% family-wise confidence level
##
## Fit: aov(formula = expenses ~ f_country * f_sex, data = tourism)
##
## $f_country
##              diff        lwr       upr      p adj
## 2-1 -14.87500000 -37.39186  7.641862 0.3184857
## 3-1 -14.97115385 -42.39662 12.454315 0.4895414
## 4-1 -23.87500000 -60.97173 13.221732 0.3417014
## 3-2  -0.09615385 -28.81176 28.619455 0.9999998
## 4-2  -9.00000000 -47.06044 29.060444 0.9272432
## 4-3  -8.90384615 -50.05897 32.251283 0.9429897
##
## $f_sex
##         diff      lwr      upr p adj
## 1-0 40.98582 26.21078 55.76085 2e-07
##
```

```
## $`f_country:f_sex`
##                  diff        lwr        upr      p adj
## 2:0-1:0   -0.06451613 -35.512881  35.38385 1.0000000
## 3:0-1:0   -6.96057348 -48.316999  34.39585 0.9995543
## 4:0-1:0  -28.11612903 -95.374673  39.14241 0.9026314
## 1:1-1:0   47.72629521  12.819158  82.63343 0.0011635
## 2:1-1:0   27.60151803 -14.517416  69.72045 0.4747743
## 3:1-1:0   46.98387097  -8.359871 102.32761 0.1603442
## 4:1-1:0   21.34101382 -37.060484  79.74251 0.9504892
## 3:0-2:0   -6.89605735 -48.252483  34.46037 0.9995809
## 4:0-2:0  -28.05161290 -95.310156  39.20693 0.9036971
## 1:1-2:0   47.79081134  12.883675  82.69795 0.0011385
## 2:1-2:0   27.66603416 -14.452900  69.78497 0.4716416
## 3:1-2:0   47.04838710  -8.295355 102.39213 0.1590688
## 4:1-2:0   21.40552995 -36.995968  79.80703 0.9496985
## 4:0-3:0  -21.15555556 -91.706807  49.39570 0.9834211
## 1:1-3:0   54.68686869  13.793403  95.58033 0.0016484
## 2:1-3:0   34.56209150 -12.637188  81.76137 0.3272164
## 3:1-3:0   53.94444444  -5.357302 113.24619 0.1035376
## 4:1-3:0   28.30158730 -33.863535  90.46671 0.8556228
## 1:1-4:0   75.84242424   8.867554 142.81729 0.0147855
## 2:1-4:0   55.71764706 -15.283266 126.71856 0.2425657
## 3:1-4:0   75.10000000  -4.461642 154.66164 0.0793909
## 4:1-4:0   49.45714286 -32.261032 131.17532 0.5790891
## 2:1-1:1  -20.12477718 -61.789225  21.53967 0.8136724
## 3:1-1:1   -0.74242424 -55.741073  54.25622 1.0000000
## 4:1-1:1  -26.38528139 -84.459859  31.68930 0.8569295
## 3:1-2:1   19.38235294 -40.453655  79.21836 0.9742725
## 4:1-2:1   -6.26050420 -68.935483  56.41447 0.9999868
## 4:1-3:1  -25.64285714 -97.872202  46.58649 0.9575119
```

```
> with(tourism, plotMeans(expenses, f_country, f_sex, error.bars="none"))
```

Plot of Means

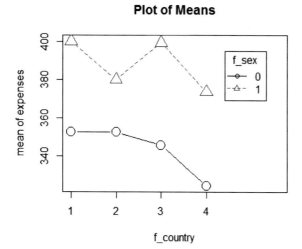

Here the reader is invited to take a look at the command structure by himself (we have abstained from testing the assumptions as this has already been shown above). We focus on the result of the analysis of variance.

The result is to be interpreted similarly to the one with the analysis of variance with one factor. The difference is, however, that we now have the two factors *f_country* and *f_sex*, a so-called interaction term *f_country:f_sex*, and the *residuals*. Again we concentrate on the columns *Sum Sq* and *Pr(>F)* in the columns. In the first column, we see the variance explained by the factors and the interaction term as well as the unexplained variance. In the second column, for both factors and interaction term, we see the p-value. We neither reject the null hypothesis for the factor *f_country* nor for the interaction term *f_country:f_sex*, but we do so for the factor *f_sex* at each level of significance specified in advance. This means that only gender has an influence on the expenses of the guests. We are at the end of our analysis and only have to look at the post hoc test for the gender variable. Here we find an average positive difference between the group of women (1) and the group of men (0) of about CHF 41, which is significant. The 95% confidence interval lies approximately between CHF 26 and CHF 56, i.e. women spend more than men do. This result can also be illustrated graphically with the help of the plot of mean.

At this point, we would like to remind you that the script is not intended to give an introduction to statistics, but the aim is to introduce R. The interested reader has to deepen his knowledge about the analysis of variance on his or her own.

6.7 Correlation Test for Metric, Ordinal and Nominal Variables

At the end of this section, we want to show you how to test relationships between variables with RStudio, if the variables are correlated or if they are independent from each other. Depending on the scale of measurement, we use different commands.

If the variables are metric, the command is as follows:

```
> with(tourism, cor.test(expenses, satisfaction, alternative="two.sided",
method="pearson"))

##
##  Pearson's product-moment correlation
##
## data:  expenses and satisfaction
## t = 6.0805, df = 148, p-value = 9.734e-09
## alternative hypothesis: true correlation is not equal to 0
## 95 percent confidence interval:
##  0.3089528 0.5667335
## sample estimates:
##       cor
## 0.4470781
```

With *with()* we first specify the dataset we want to use. With *cor.test()* we specify that we would like to test for a correlation. Within the brackets, we write the variables first, with *alternative = "two.sided"* we tell RStudio that we test two-sided. Alternatively, we could enter *alternative = "greater"* for a positive relationship or *alternative = "less"* for a negative relationship. With *method = "pearson"* we tell RStudio that we are testing for a correlation with metric variables.

When we look at the result, we see the two variables *expenses* and *satisfaction*. Then we get the calculated t-value, the number of degrees of freedom (df) and the p-value. The p-value is very small, so we reject the null hypothesis for each specified significance level (10%, 5%, 1%), i.e. we assume that the alternative hypothesis applies. The alternative hypothesis is specified in the next line. This is followed by the 95% confidence interval and the correlation coefficient of the sample.

The command for ordinal data is almost the same, we just specify *method = "spearman"*:

```
> with(tourism, cor.test(recommendation, education, alternative="two.sided",
method="spearman", exact=FALSE))

##
##   Spearman's rank correlation rho
##
## data:   recommendation and education
## S = 544000, p-value = 0.6898
## alternative hypothesis: true rho is not equal to 0
## sample estimates:
##          rho
## 0.03285428
```

Let's take a look at the result. First, we get the used variables *recommendation* and *education*. After that, we receive the test statistics (S) and the p-value. The p-value is 0.6898. Of course, we do not reject the null hypothesis and assume that there is no relationship between the two variables. In addition to the p-value, we also see the alternative hypothesis and the correlation coefficient calculated for the sample.

If we wish to test for a relationship between nominal variables, we first have to tell RStudio that the variables are factors, generate the variables and add them to the dataset. Then we can perform the chi-squared independence test. The command structure is as follows:

```
> f_skiholiday <- as.factor(tourism$skiholiday)
> f_sex <- as.factor(tourism$sex)
> tourism <- cbind(tourism, f_skiholiday, f_sex)


> table(tourism$f_skiholiday, tourism$f_sex)

##
##       0  1
## 0 40 17
## 1 45 48

> chisq.test(tourism$f_skiholiday, tourism$f_sex)

##
##   Pearson's Chi-squared test with Yates' continuity correction
##
## data:   tourism$f_skiholiday and tourism$f_sex
## X-squared = 5.9738, df = 1, p-value = 0.01452
```

Let's take a quick look at the command structure. In the first three lines, we define the factors and add them to the data set. With the fourth line and the command *table()* we instruct RStudio to output the observed frequencies. In the last line we find the command *chisq.test()*. With this, we perform the chi-square independence test for the two specified variables *f_skiholiday* and *f_sex*.

As a result we see that we have 40 men (column variable = 0) who don't want to spend any more ski holidays in our destination (row variable = 0) and 45 men who would spend ski holidays in our destination again (row variable = 1). For women (column variable = 1), the ratio is 17 to 48. We also see the Pearson Chi-square test. In the last line of the test result the calculated chi-squared value, the number of degrees of freedom (df), and the p-value are displayed. The p-value is 1.45%. Depending on the level of significance, we have specified in advance, we reject the null hypothesis or not. Let's assume that we test at the 5% significance level, then we reject the null hypothesis and assume a correlation between the two variables. Gender and the question of whether another skiing holiday will be spent are not independent of each other. If we now look at the observed frequencies, we see that women are more likely going to spend their holidays in our destination again than men.

Table 6.1 R commands learned in this chapter

Command	Notes/examples
aov()	....
	....
cor.test()	....
	....
chisq.test()	....
	....
leveneTest()	....
	....
plotMeans()	....
	....
tapply()	....
	....
t.test()	....
	....
TukeyHSD()	....
	....
wilcox.test()	....
	....

Note You can take notes or write down examples for their usage on the right side

6.8   Time to Try

6.1. The tourism manager of the destination tries to figure out whether the satisfaction of guests has increased over the past years. He knows that a few years ago, the same survey was conducted and the overall satisfaction was only 60 points. Use the dataset tourism.xlsx and test at the 5% significance level whether the satisfaction with the ski resort has increased.

6.2. The tourism manager wants to find out whether women spend more money per day than men do. Test this issue at the 10% level with the help of the dataset tourism. xlsx.

6.3. You would like to find out whether women and men differ in their education. Check at the significance level of 5% with the correct test.

6.4. Use the dataset schoolbreak.xlsx and use it to build a reduced dataset named schoolbreak_red with only eight observations. Test whether the difference between weight before and weight after is normally distributed. Run the t-test for dependent samples and the Wilcoxon signed-rank test and decide at the 5% significance level whether active school breaks help to reduce weight.

6.5. As tourism manager you would like to find out whether there is a difference in spending behavior with regard to the education level of guests. Use the dataset tourism.xlsx and one-way ANOVA to find out at the 10% significance level.

6.6. You extend your analysis of task 6.5 and you would like to include the gender of guests as well. In doing so you run a two-way ANOVA with the factors education and sex. Find out at the 10% significance level whether there is a difference in spending behavior (dataset tourism.xlsx).

6.7. Test for a correlation between diversity and satisfaction, waitingtime and satisfaction, safety and satisfaction, quality and satisfaction (dataset tourism.xlsx). Use the correct correlation coefficient and use a significant level of 5%. Do not forget to evaluate the scatterplot first.

6.8. Use the dataset tourism.xlsx and calculate the correct correlation coefficient for the variables *education* and *accommodation* of the dataset tourism.xlsx. Test at a significance level of 5%.

6.9. Use the dataset tourism.xlsx and check whether there is a relationship between the variables *accommodation* and *skiholiday*. Test at the significance level of 10%.

6.10. Do the same as in task 6.9 for the variables *country* and *skiholiday*.

Linear Regression with RStudio

7

In this chapter first, we will discuss how to perform a regression analysis with RStudio and second how to check the assumptions of the linear regression model. Before we start, we clear our memory and reload the data.

```
> rm(list=ls())
> library(readxl())
> tourism <- read_excel("D:/serie_r/tourismus/tourism.xlsx")
```

7.1 Simple and Multivariate Linear Regression

Let's start with a short introduction to linear regression. The principle of linear regression is very simple. We look for an equation that describes the linear relationship between a dependent variable and one or more independent variables. When we have only one independent variable we speak of a simple linear regression, when we have more than one independent variable than we speak of a multivariate or multiple regression. As a result of the linear regression, the relationship is described. We get information about how strong the relationship is and we can make predictions. For example, we can predict what happens to the dependent variable when the independent variable has a certain value.

Let's illustrate this with a small example. Let's assume that there is a relationship between the daily expenses per guest and the guest's satisfaction with the ski area. At a first glance, we can analyze and describe this relationship with a scatterplot and draw the scatterplot between the satisfaction and the daily expenses.

```
> plot(tourism$satisfaction, tourism$expenses)
```

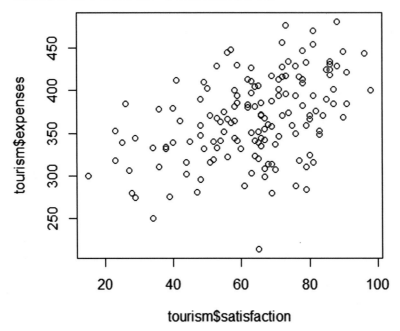

With the linear regression, we look for a straight line and its equation that best fits to the cloud of points.

Before we proceed with the example, we would like to make a few important remarks. First, as the name linear regression implies, the relationship must be described by a straight line. Second, it must be a causal relationship. The independent variable(s) must influence the dependent variable, not the other way around. In our example, we must be able to explain theoretically that guest satisfaction affects the level of spending. The relationship must not run in the other direction (Fig. 7.1: Causal relationship between dependent and independent variable).

More generally, this can be illustrated as follows (Fig. 7.2: Causal relationships in regression). A theoretical model tells us that one or more independent variables influence a dependent variable. The dependent variable is metric, the independent variables are also metric, or so-called dummy variables, variables with two characteristics (0 and 1).

At this point, an important note has to be made: causality is always a theoretical issue, never a matter of data. This means that regression always requires a theoretically justified approach!

Fig. 7.1 Causal relationship between dependent and independent variable

$$Y \longleftarrow X_1, X_2, ..., X_J$$

dependent variable   independent variable(s)

(metric)   (metric or dummy-variables)

Fig. 7.2 Causal relationships in regression

Now let's go back to the connection between the level of expenses and the satisfaction of the guests. Mathematically we can describe this as follows:

$$expenses_i = \beta_0 + \beta_1 * satisfaction_i + u_i$$

The expenses of a guest i can be described by the coefficients β_0 and β_1 multiplied by the satisfaction of the guest plus a remaining term u_i. The coefficient β_0 is the intercept of the straight line and the coefficient β_1 is the slope of the straight line, the so-called marginal effect, the change of the dependent variable when the independent variable increases by one unit. u_i is called the "error term" and it is the random discrepancy between the observed expenses and the estimated expenses.

The notation shown above with the coefficients β_0, β_1 and u_i describes the relationship within the population and is the true function. In fact, we do not know the true function; we estimate the straight line and its coefficients from our sample data. The formula of the estimated straight line is as follows:

$$\widehat{expenses}_i = b_0 + b_1 * satisfaction_i$$

b_0 is the intercept estimated by our data and b_1 is the slope of the straight line, i.e. the change when satisfaction increases by one point. The coefficients b_0 and b_1 are the estimated coefficients based on the sample. If everything went well, then these coefficients are identical with the true coefficients of the population (more on this in Sect. 7.2). Please also note the small hat above the dependent variable. The hat indicates that $\hat{y}_i$ is the estimated expense of guest i. More generally, we write:

$$\hat{Y} = b_0 + b_1 * X$$

If we have one independent variable, we speak of a simple linear regression. If we include several independent variables in the model, it is a multivariate regression. We estimate the influence of the independent variables simultaneously, and we get the following formula:

$$\hat{Y} = b_0 + b_1 * X_1 + b_2 * X_2 + \cdots + b_J * X_J$$

Before we continue with the example, a note on the estimation techniques is necessary. As we said, we estimate the regression coefficients using sample data. The method typically used is the "ordinary least squares estimator OLS". The ordinary least squares estimation technique is one of the best estimation methods we know, as long as certain assumptions are met (for more information, see Sect. 7.2).

Starting point of OLS is the difference between each observed value y_i and each estimated value $\hat{y}_i$. This difference is called "residual" and is indicated by the letter e. The difference is accordingly $e_i = y_i - \hat{y}_i$, which we can calculate for each observation, i.e. each guest.

The objective of the ordinary least squares estimation technique is now to minimize the sum of the differences for all observations. However, the sum cannot be minimized, because it is always zero. With an optimal straight line, we have as many positive deviations as negative ones. But the sum of the squared differences $\sum e_i^2 = \sum(y_i - \hat{y}_i)^2$ can be minimized. We get the minimum if we replace $\hat{y}_i$ with the regression line, thus obtaining $\sum e_i^2 = \sum(y_i - (b_0 + b_1 * x_i))^2$, and take the partial derivatives of this expression with respect to b_0 and b_1. At this point, the interested reader himself is required to broaden his knowledge. But what should have become clear is why the method is called "ordinary least squares" estimation technique or OLS.

Now let's start with the linear regression with RStudio. The command that requests a linear regression is as follows:

```
> RegModel.1 <- lm(expenses~satisfaction, data=tourism)
> summary(RegModel.1)

##
## Call:
## lm(formula = expenses ~ satisfaction, data = tourism)
##
## Residuals:
##      Min       1Q   Median       3Q      Max
## -155.621  -28.419    1.417   31.821   97.021
##
## Coefficients:
##               Estimate Std. Error t value Pr(>|t|)
## (Intercept)   285.4582    14.0361   20.34  < 2e-16 ***
## satisfaction    1.2948     0.2129    6.08 9.73e-09 ***
## ---
## Signif. codes:  0 '***' 0.001 '**' 0.01 '*' 0.05 '.' 0.1 ' ' 1
##
## Residual standard error: 44.76 on 148 degrees of freedom
## Multiple R-squared:  0.1999, Adjusted R-squared:  0.1945
## F-statistic: 36.97 on 1 and 148 DF,  p-value: 9.734e-09
```

With the assignment operator "<-" we specify under which name we save the results of the regression analysis. In our case, we call the object *RegModel.1*. We could also choose another name. Here the idea is quite simple, during our analysis we will usually calculate several models. We name each of these models, the first *RegModel.1*, the second *RegModel.2* and so on, so that we do not simply overwrite the results. So we can use the results again, if necessary. The command for the regression analysis is *lm()*. lm stands for linear model. Within the brackets, we first specify the dependent variable. The tilde " $\sim$ " separates the independent variables from the dependent variable, at the end we specify which dataset is used.

Let's turn to the result. After entering the command *summary()*, we get the result of the regression analysis in four output blocks. Let's take a quick look at them one after the other.

In the output block *Call* the specification of the calculated model is specified, the command, the dependent and independent variables as well as the dataset.

In the output block *Residuals* we get a quick overview of the size and distribution of the residuals, the minimum, the 1st quartile, the median, the 3rd quartile and the maximum. We will see later that the residuals should be normally distributed, i.e. the median should be approximately zero, furthermore our model is better the smaller the residuals are, which is the difference between observed dependent variable values and estimated values.

Let's now focus on the *model fit* block. In the model fit block, we see the F-statistic in the last line, which is our test statistic. We can take the value of the F-statistic and compare it with the value of the F-distribution table or we can look at the p-value. The null hypothesis of the regression model is: The model does not explain the dependent variable. The corresponding alternative hypothesis is: The model explains the dependent variable. If now the p-value is small, we reject the null hypothesis and proceed with the alternative hypothesis. The significance level is defined in advance of the analysis and is usually 1% ($\alpha = 0.01$). In our example, the p-value is much smaller at 0.000000009734 (p-value: 9.734e-09), i.e. we reject the null hypothesis and assume that the model explains the dependent variable. In the line above, we see the R^2 (Multiple R-squared). The R^2 shows how much variation of the dependent variable is explained by the model. Here it is around 20%, i.e. 20% of the variation of the expenses is explained by the model, 80% remain unexplained. Additionally we see the adjusted-R^2 (adjusted R-squared). We use the adjusted-R^2 when we calculate more models with a different number of independent variables. We check the adjusted-R^2 to evaluate which of these models is the better model. Usually, we take and interpret the model with the highest adjusted-R^2. Finally, the standard error of the residuals is displayed. It is also a measure for the quality of the regression model. The standard error can be interpreted as the average deviation of the estimated y-values $\hat{Y}$ from the true values. The value is displayed in the unit of the dependent variable and is therefore easy to interpret. In our example, we are on average 44.76 Swiss francs off with our model. After this, the degrees of freedom (df) are displayed. The number shows on how many observations minus the estimated parameters the model is based. In our case we have 148 degrees of freedom, n = 150 observations minus two estimated parameters (restrictions), b_0 and b_1.

Let's now turn to the *coefficient* block. This is a matrix and we get the coefficients of the regression model (Estimate), the standard error (Std. Error), the t-value and the p-value (Pr(>|t|)). We have a null hypothesis and an alternative hypothesis for each variable. The null hypothesis is: The variable does not contribute to the explanation of the dependent variable. The alternative hypothesis is: The variable contributes to the explanation of the dependent variable.

The coefficients are our parameters for the respective variables of the regression model. The standard error tells us how large the average error is for the coefficient we estimated based on the sample. In principle, we could have drawn a different sample and obtained a different value for the coefficient. The t-value is our test statistic and the Pr(>|t|) is used to reject the null hypothesis or not. The significance level, is also defined in advance of the analysis and is usually 10%, 5% or 1% ($\alpha = 0.1, 0.05, 0.01$). In our case, the p-value for both the coefficient and the independent variable is very small, i.e. we would reject the null hypothesis for each specified level of significance and assume that the variables contribute to the explanation of the model.

With this, we can finally establish and interpret our regression line:

$$\hat{Y} = 285.46 + 1.29 * X$$

The coefficient of the intercept tells us how much money the guests spend per day if they have a satisfaction of zero (X = 0). The coefficient for the independent variable *satisfaction* tells us what happens when satisfaction increases by one point. The expenses increase by 1.29 Swiss francs. Now we can make an additional prediction, e.g. calculate the expenses for a satisfaction of 86 points. We can make the forecast directly in the console by entering the following information:

```
> 285.46+1.29*86
```

```
## [1] 396.4
```

With a satisfaction of 86 points, the daily expenses are on average 396.4 Swiss francs.

In many cases, one independent variable will not be sufficient to describe a dependent variable. Usually, a dependent variable is influenced by several factors. The regression model must include all factors that have an influence on the dependent variable. We speak of a multivariate or multiple regression model.

Assume that the expenses do not only depend on satisfaction, but theoretically also on age, length of stay and gender. Our regression model is then specified as follows:

$$\hat{Y} = b_0 + b_1 * X_1 + b_2 * X_2 + b_3 * X_3 + b_4 * X_4$$

or

$$\hat{Y} = b_0 + b_1 * age + b_2 * stay + b_3 * sex + b_4 * satisfaction$$

We estimate the model as above, except that we now have to add several independent variables. The command is as follows (we have specified a new name to avoid overwriting the previous result):

```
> RegModel.2 <- lm(expenses~age+stay+sex+satisfaction, data=tourism)
> summary(RegModel.2)

##
## Call:
## lm(formula = expenses ~ age + stay + sex + satisfaction,
##      data = tourism)
##
## Residuals:
##      Min       1Q  Median       3Q      Max
## -84.713 -20.499   1.323   18.550   83.879
##
## Coefficients:
##                 Estimate Std. Error t value Pr(>|t|)
## (Intercept)     221.0441    14.5154  15.228  < 2e-16 ***
## age               2.5009     0.2226  11.236  < 2e-16 ***
## stay             -0.8873     1.3076  -0.679  0.49847
## sex              14.9424     5.7074   2.618  0.00978 **
## satisfaction      0.4066     0.1675   2.428  0.01642 *
## ---
## Signif. codes:  0 '***' 0.001 '**' 0.01 '*' 0.05 '.' 0.1 ' ' 1
##
## Residual standard error: 31.28 on 145 degrees of freedom
## Multiple R-squared:  0.6172, Adjusted R-squared:  0.6067
## F-statistic: 58.45 on 4 and 145 DF,  p-value: < 2.2e-16
```

The interpretation is similar. We restrict ourselves to the most important points.

The line with the F statistics shows that the model explains the dependent variable. The p-value is very small, in other words the model is significant. In total, 61.7% of the variation of the dependent variable is explained by the model. 38.3% of the variation of Y is not explained.

Assume that we have set the significance level at 5% prior to the analysis ($\alpha = 0.05$), then the intercept, age, gender and satisfaction are significant. Here the p-value is below the 5% level. The length of stay, on the other hand, has no significant influence on the dependent variable. The p-value for the length of stay is 49.85%.

The regression model is as follows:

$$\hat{Y} = 221.04 + 2.50 * age - 0.89 * stay + 14.94 * sex + 0.41 * satisfaction$$

Let's interpret the regression model. We notice that we have included the coefficients and independent variables, regardless of whether they were significant or not. We have to do this because we estimated the model with all variables. But only the significant variables have to be interpreted. For the non-significant variables, we assume that they have no influence on the dependent variable in the population.

We see that the average expenses are 221 Swiss francs, if all variables are zero. This is more or less nonsense, because a baby will hardly spend any money. With each year of ageing, the level of expenditure increases by 2.50 Swiss francs, if the other variables do not change and with each point of increasing satisfaction the expenditure raises by 0.41 Swiss francs. In addition, the average level of expenditure of women is CHF 14.94 higher than of men. Gender is a so-called dummy variable with two characteristics (0 = man and 1 = woman). If the variable changes from 0 to 1, then the coefficient of a dummy variable indicates this effect. We only interpret the significant variables.

What does "controlling for" mean?

At this point, it is worth explaining what the term "controlling for" means. We have seen that normally several factors are necessary to explain the dependent variable. Usually, however, we are only interested in one factor, not in all. But to know what happens when we change the factor of interest, we need to control for all other factors that have an influence on the dependent variable. We have to calculate out the influence of the other variables on the dependent variable, so that the coefficient of the variable of interest contains only its influence. We estimate the influence of the variable of interest on the dependent variable "controlling for" the influence of all other independent variables.

Now there is something very important, which makes our work much easier. We mentioned above that we need a causal relationship for the independent variables, such as the independent variable influences the dependent variable. If this is not the case, the regression model only calculates a correlation between the independent variable and the dependent variable. We cannot say what will happen if the independent variable changes. However, in order to establish a causal relationship, we need to know the literature for the variable. This means that we have to read the literature for the independent variables, which we would like to interpret. This can be very time-consuming. However, we don't have to do this for the control variables. If we don't want to make a forecast for them, we don't need a causal relationship, in the sense that the independent variable influences the dependent variable. We only need the control variables to determine the real influence of the variable of interest.

We now want to expand our model once more. So far, we have allowed metric independent variables and nominal variables with two characteristics in the model. This is correct and will remain. But what do we do if we want or have to include nominal variables with more than two characteristics or ordinal variables in the model? In such a case, we can build so-called "dummy variables" from a nominal or ordinal variable with more than two characteristics. Each nominal variable with more than two characteristics and each ordinal variable can be described by a set of dummy variables with the two characteristics "0" and "1". For this, we need the same number of dummy variables as the original variable has characteristics. For example, if we have a nominal or ordinal variable with four characteristics, we need four dummy variables. If we have a variable with five characteristics, we form five dummy variables and so on. Let's demonstrate this with an example.

Let's take the variable country from our dataset. It is a nominal variable with four characteristics. Our guests come from either Switzerland, Germany, Austria or other countries. Our coding is Switzerland = 1, Germany = 2, Austria = 3 and other countries = 4. If we would like to describe this variable by dummy variables, we need four dummy variables, one for Switzerland, one for Germany, one for Austria and one for the other countries. In the respective dummy variable, we assign a "1", if the fact applies, otherwise a "0". Let's take the dummy variable for Switzerland. We assign a "1" if the guest comes from Switzerland, a "0" if the guest comes from another country. Let's have a look at the example now.

Dummy variables can be created easily with RStudio once we have installed the necessary package. The package we can use is called *"fastDummies"*. First, we install the package with the command required for this purpose:

```
> install.packages("fastDummies")
```

Then we activate the package and run the command. We specify the name of the dataset in which we want to generate the dummies, followed by the assignment operator <-. The function itself is *dummy_cols()*. Within the brackets, we specify the dataset again with the variable(s) we would like to encode to dummies. With *select_columns = c()* we select the relevant column. We get four new dummy variables with the column headings *"country_1"*, *"country_2"*, *"country_3"* and *"country_4"*. If we are not satisfied with that, but want to name the columns with the respective country, we do this with the following four commands *colnames()*. Everybody is invited to look at the function himself or herself.

```
> library(fastDummies)
> tourism <- dummy_cols(tourism, select_columns =  c("country"))
> colnames(tourism)[colnames(tourism)=="country_1"] <- "CH"
> colnames(tourism)[colnames(tourism)=="country_2"] <- "GER"
> colnames(tourism)[colnames(tourism)=="country_3"] <- "AUT"
> colnames(tourism)[colnames(tourism)=="country_4"] <- "Oth"
```

Let's take a look at the result. In the data set, we have four new columns with the headings CH, GER, AUT and Oth (Fig. 7.3: Creating dummy variables).

These dummy variables include the information of the variable *country*. A guest who comes from Switzerland has a "1" for Switzerland and a "0" for the other countries, and so on. Thus, each guest is assigned uniquely to a country. The information of the variable *country*, meaning which guest comes from which country, is now also included in the four generated dummy variables.

As already mentioned several times, we can include nominal variables with two values in the regression model, i.e. the generated dummy variables can be integrated in the model. We only have to consider one further aspect. For reasons that we do not explain in this simple script, we cannot include all four dummy variables, but only three of them. More generally, if we have generated k dummy variables from one variable, we can only include $k - 1$ dummy variables in the model. The dummy variable we omit (it doesn't matter which one) is our reference category. The coefficients of the included dummy variables tell us how they differ in comparison to the reference category. As an example, we include three of the generated dummy variables in the regression model. We want to

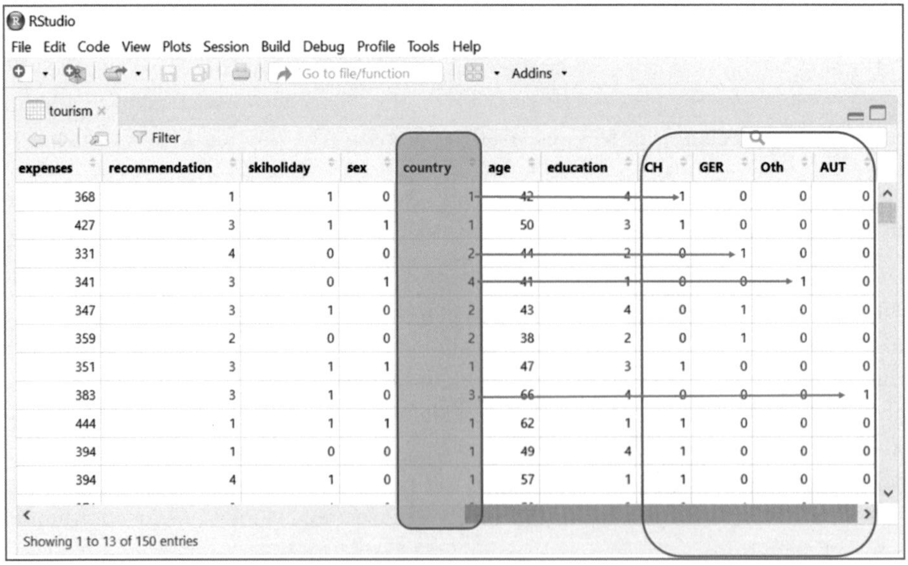

expenses	recommendation	skiholiday	sex	country	age	education	CH	GER	Oth	AUT
368	1	1	0	1	42	4	1	0	0	0
427	3	1	1	1	50	3	1	0	0	0
331	4	0	0	2	44	2	0	1	0	0
341	3	0	1	4	41	1	0	0	1	0
347	3	1	0	2	43	4	0	1	0	0
359	2	0	0	2	38	2	0	1	0	0
351	3	1	1	1	47	3	1	0	0	0
383	3	1	0	3	66	4	0	0	0	1
444	1	1	1	1	62	1	1	0	0	0
394	1	0	0	1	49	4	1	0	0	0
394	4	1	0	1	57	1	1	0	0	0

Fig. 7.3 Creating dummy variables

examine whether guests from different countries differ in their spending behavior. We estimate the following regression model:

$$\hat{Y} = b_0 + b_1 * age + b_2 * stay + b_3 * sex + b_4 * satisfaction$$
$$+ b_5 * GER + b_6 * AUT + b_7 * Oth$$

```
> RegModel.3 <- lm(expenses~age+stay+sex+satisfaction+GER+AUT+Oth,
data=tourism)
> summary(RegModel.3)

##
## Call:
## lm(formula = expenses ~ age + stay + sex + satisfaction +
##     GER + AUT + Oth, data = tourism)
##
## Residuals:
##    Min     1Q Median     3Q    Max
## -80.69 -20.31  -1.00  18.30  80.48
##
## Coefficients:
##               Estimate Std. Error t value Pr(>|t|)
## (Intercept)   226.8513    14.5250  15.618  < 2e-16 ***
## age             2.5003     0.2180  11.468  < 2e-16 ***
## stay           -0.8826     1.3184  -0.669  0.50427
## sex            14.3896     5.6950   2.527  0.01261 *
## satisfaction    0.4246     0.1646   2.579  0.01093 *
## GER           -10.6893     5.9543  -1.795  0.07474 .
## AUT            -6.3674     7.3016  -0.872  0.38465
## Oth           -27.3146     9.6535  -2.830  0.00534 **
## ---
## Signif. codes:  0 '***' 0.001 '**' 0.01 '*' 0.05 '.' 0.1 ' ' 1
##
## Residual standard error: 30.62 on 142 degrees of freedom
## Multiple R-squared:  0.6408, Adjusted R-squared:  0.623
## F-statistic: 36.18 on 7 and 142 DF,  p-value: < 2.2e-16
```

First, we look at the model fit block. We see that the model explains the spending behavior of the guests. Based on the F statistics, we reject the null hypothesis that the model explains nothing. The R^2 shows us that we explain 64.1% of the variation of the expenses behaviour. Let's assume that we check the independent variables at the 5% significance level. Significant are hence the intercept, age, sex, satisfaction and the variable other countries. The regression equation is:

$$\hat{Y} = 226.9 + 2.50 * age - 0.88 * stay + 14.39 * sex +$$
$$0.43 * satisfaction - 10.69 * GER - 6.37 * AUT - 27.32 * Oth$$

We see that 226.9 Swiss francs are spent when all independent variables are zero. If the age rises by one year, the expenditure rises by 2.5 Swiss francs. Women spend 14.4 Swiss francs more than men do. If satisfaction rises by one point, the expenses increase by 0.4 Swiss francs. For the dummy variables, only the coefficient for the other countries "Oth" is significant. This means that guests from other countries spend 27.3 Swiss francs less than guests from Switzerland (reference category). German and Austrian guests do not differ from Swiss guests in their spending behavior.

For a first simple introduction into regression analysis with RStudio we want to leave it at that. In the next chapter, we will turn to the evaluation of the estimated regression model, whether the results are trustworthy or the model should be improved.

7.2   Regression Diagnostic with RStudio

In the subchapter above, we calculated and interpreted several regression models. Expenses were estimated as a function of age (RegModel.1), of age, length of stay, gender and satisfaction (RegModel.2) and of age, length of stay, gender, satisfaction and country of origin (RegModel.3).

The estimation technique we have used is the method of ordinary least squares OLS. This method is one of the best we know, as long as certain assumptions are fulfilled. The assumptions are:

1. the model contains all relevant and no irrelevant variables and the relationship between the independent variables and the dependent variable is linear,
2. the error term is uncorrelated with the independent variables,
3. the deviation of observed value and estimated value, the error term $e_i = y_i - \hat{y}_i$ has an expected value of zero,
4. the variance of the error term e_i is constant (no heteroscedasticity),
5. the independent variables are not correlated to each other (no multicollinearity),
6. the error term is normally distributed.

Furthermore, there is an assumption that states that the error term should not be correlated with itself (autocorrelation). Autocorrelation however is basically a matter of time series analysis and not a matter of cross section analysis, which is shown here.

If the assumptions are fulfilled, the method of ordinary least squares OLS is the best, unbiased and most efficient estimator we know, i.e. the β-coefficients are estimated unbiased and as best as possible. However, the results of regression analyses are strongly influenced by outliers or influential observations as well. Hence, in addition to testing the assumptions it is necessary to identify outliers and influential observations and think about how to deal with them.

Overall, we must check the regression model to see whether the assumptions are fulfilled and whether there are any influential observations. Hence, after estimating the regression model we carry out the so-called regression diagnostics.

We want to discuss regression diagnostics using the third model (RegModel.3). Before we continue, a short side note has to be made. We only discuss graphical methods. Of course, there are also test procedures that can be applied additionally. At this point, we refer to the further literature.

Before we begin, we clear the working memory, load the data and re-estimate the model. In addition, we load or install *ggplot2* and *ggfortify*, two packages that are helpful for the graphical regression diagnostic.

```
> rm(list=ls())
> library(readxl())
> tourism <- read_excel("D:/serie_r/tourismus/tourism.xlsx")
> library(fastDummies)
> tourism <- dummy_cols(tourism, select_columns = c("country"))
> colnames(tourism)[colnames(tourism)=="country_1"] <- "CH"
> colnames(tourism)[colnames(tourism)=="country_2"] <- "GER"
> colnames(tourism)[colnames(tourism)=="country_3"] <- "AUT"
> colnames(tourism)[colnames(tourism)=="country_4"] <- "Oth"
> RegModel.3 <- lm(expenses~age+stay+sex+satisfaction+GER+AUT+
  Oth, data=tourism)
> summary(RegModel.3)
```

```
##
## Call:
## lm(formula = expenses ~ age + stay + sex + satisfaction +
##      GER + AUT + Oth, data = tourism)
##
## Residuals:
##     Min     1Q Median     3Q     Max
## -80.69 -20.31  -1.00  18.30  80.48
##
## Coefficients:
##                 Estimate Std. Error t value Pr(>|t|)
## (Intercept)     226.8513    14.5250  15.618  < 2e-16 ***
## age               2.5003     0.2180  11.468  < 2e-16 ***
## stay             -0.8826     1.3184  -0.669  0.50427
## sex              14.3896     5.6950   2.527  0.01261 *
## satisfaction      0.4246     0.1646   2.579  0.01093 *
## GER             -10.6893     5.9543  -1.795  0.07474 .
## AUT              -6.3674     7.3016  -0.872  0.38465
## Oth             -27.3146     9.6535  -2.830  0.00534 **
## ---
## Signif. codes:  0 '***' 0.001 '**' 0.01 '*' 0.05 '.' 0.1 ' ' 1
##
## Residual standard error: 30.62 on 142 degrees of freedom
## Multiple R-squared:  0.6408, Adjusted R-squared:  0.623
## F-statistic: 36.18 on 7 and 142 DF,  p-value: < 2.2e-16
```

The first assumption is fulfilled when we have theoretically worked well and have established a model that includes all relevant factors influencing the dependent variable. In addition, as described above, the relationship between dependent variables and independent variables must be represented by a straight line (see Sect. 7.1). A simple check can be done with the help of the scatterplot between the dependent variable and the metric independent variables. For the dummies this is not necessary because there can be no linear relationship. Typically, we do this before we estimate the model.

To receive the scatterplot for all of the relevant variables we can use the command *pairs()* to plot all of the scatterplot at once or we use simply the command *plot()* to produce the necessary scatterplots one after another.

```
pairs(~expenses+age+stay+satisfaction, data=tourism, main="Simple  Scatterplot
Matrix")
```

Simple Scatterplot Matrix

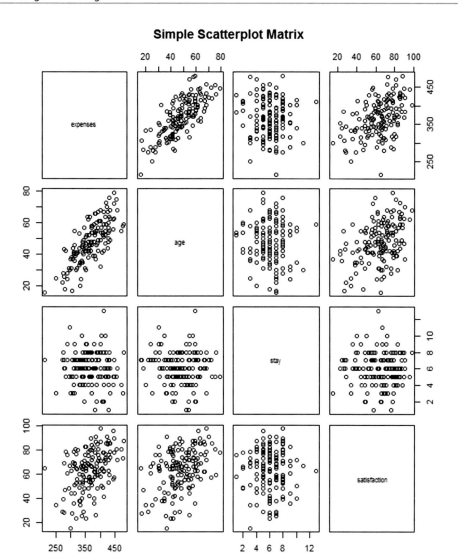

To evaluate we focus on either on the horizontal row or the vertical column where the dependent variable shows up. Then we look at the scatterplots, when we detect a linear relationship or no relationship at all the assumption is not violated. If we have a non-linear relationship then we have to think about solutions for the violation of the assumption.

The second assumption demands no correlation between the residual and the independent variables. Again, we can use the scatterplot matrix to check the assumption. Here we plot the residuals against the independent variables. To do so, it is best that we add the residuals to our dataset first (we can do the same with the fitted values and so on). Afterwards we can produce the scatterplot matrix as we did above.

```
tourism <- cbind(tourism, RegModel.3$residuals)

pairs(~RegModel.3$residuals+age+stay+satisfaction, data=tourism,   main="Simple
Scatterplot Matrix")
```

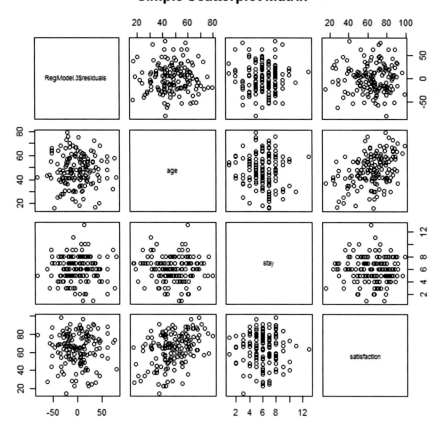

When we find no relationship between the residuals and the metric independent variables, the assumption is not violated. Again, we focus either on the row or on the column of the residuals.

The other four assumptions, assumption three, four, five and six, as well as unusual and influential observations can easily graphically be checked with the command *autoplot()*. To use it we first have to install and load the packages *ggplot* and *ggfortify*.

```
> library(ggplot2)
> library(ggfortify)
> autoplot(RegModel.3)
```

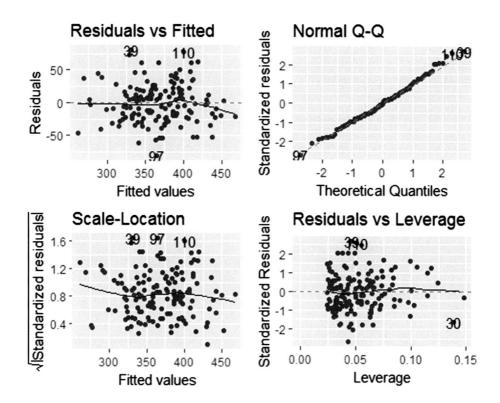

We get four plots in one graph in the environment window. Now we will look more closely at these plots.

The *Residuals vs Fitted* plot at the top left, plots the residuals e_i on the Y-axis and the estimated values $\hat{Y}$ on the X-axis. A correctly specified model equation requires that the residuals have the expected value of zero, globally and locally. The line in the plot displays the moving average of the residuals, i.e. if the line lies on the zero line and is more or less straight, then the assumption is fulfilled. At the same time, we get an impression whether the residuals have the same variance over the estimation range. This is the case when the residuals scatter over the entire estimation range in the same bandwidth. No pattern, e.g. in the form of a funnel, should be visible. The *Scale-Location* plot described as follows is also used to test whether the variance is the same.

In the *Scale-Location* plot at the bottom left, we see the root of the standardized residuals on the Y-axis. With this plot we test, as mentioned above, whether the residuals have the same variance over the estimation range (we speak of homoscedasticity). If this is given, the residuals should be independent of the estimated values, so we should see a more or less horizontal line with evenly and randomly scattering points around it. A typical test which is used to test for heterosecasticity is the Breusch-Pagan test, which should be introduced here additionally. The test is included in the package *olsrr* and can be performed by the command *ols_test_breusch_pagan()*.

```
library(olsrr)
ols_test_breusch_pagan(RegModel.3)

##
##  Breusch Pagan Test for Heteroskedasticity
##  -----------------------------------------
##  Ho: the variance is constant
##  Ha: the variance is not constant
##
##                  Data
##  ------------------------------------
##  Response : expenses
##  Variables: fitted values of expenses
##
##            Test Summary
##  ---------------------------
##  DF            =    1
##  Chi2          =    0.05130457
##  Prob > Chi2   =    0.8208086
```

Back to the *autoplot()* output. The *Normal Q-Q* plot checks whether the residuals are normally distributed. We already discussed this. The assumption is not violated when the residuals are close to the dotted line. Additionally we can use the tests discussed in Chap. 5.

With the last plot, the *Residual vs. Leverage* plot, we check for influential observations. Influential observations are observations that have on the one hand large residuals and on the other hand strongly influence the coefficients, i.e. have a large leverage effect. So we are looking for values that are in the upper right corner of the graph and its lower right corner. In our case, the observation with the number 30 would be most likely such an influential observation. However, the standardized residual with slightly more than 1.5 standard deviations is not very large. If we had influential observations, we would see a dotted line in the plot with the label Cook's distance. Influential observations would then be observations that lie above or below this line.

Last but not least we want to take a short look at the assumption that the independent variables should not be correlated with each other (we speak of no multicollinearity). The easiest way to check this is with the correlation matrix for metric data. Strictly speaking, we do not test for multicollinearity but only for collinearity (bivariate correlation between two variables).

```
> cor(tourism[,c("age", "stay", "satisfaction")])

##                       age         stay  satisfaction
## age           1.00000000 -0.02947485    0.39122036
## stay         -0.02947485  1.00000000    0.01577979
## satisfaction  0.39122036  0.01577979    1.00000000
```

Another possibility to test for multicollinearity would be the variance inflation factor, which is obtained by the command *vif()* from the package *RcmdrMisc*.

```
library(RcmdrMisc)
vif(RegModel.3)

##    age         stay        sex      satisfaction    GER         AUT

## 1.259927    1.065172    1.273931    1.277319    1.234030    1.221954

##     Oth

## 1.097120
```

If we find a variance inflation factor larger than 4, then we have for sure a problem with multicollinearity and should deal with it.

If we find out with the help of these simple methods, that assumptions are violated, you should definitely continue working on your model and trying to find solutions before interpreting the regression model. We either consult further literature (see Chap. 9) or ask a statistician.

That's all we want to do in this simple script. The script is not meant to be a statistic textbook, but a first introduction to R.

Table 7.1 R commands learned in this chapter

Command	Notes/examples
autoplot()	....
colnames()	....
dummy_cols()	....
lm()	....
ols_test_breusch_pagan()	....
pairs()	....
summary()	....
vif()	....

Note You can take notes or write down examples for their usage on the right side

7.3   Time to Try

7.1. You would like to find out whether the age of the guests influence the amount spent per day (dataset tourism.xlsx). Run a linear regression with *age* as independent and *expenses* as dependent variable. Check and evaluate the assumptions.

7.2. You are not really satisfied with you result, thinking that one independent variable may not be sufficient to explain daily spending. You go more into detail and you assume that beside the *age*, the *satisfaction* with the ski resort and length of the *stay* may also have an influence on daily spending (dataset tourism.xlsx). You rerun the regression and check the assumptions again.

7.3. You rethink the whole thing and look at your data. The measurement of the expenses attract your attention. It is measured including costs of overnight stays. Hence, you think about including the accommodation variable as well. Reestimate your model und check the assumptions again.

7.4. Again, you think that you can improve the regression by adding the variable sex. Redo everything from task 3.

Further Reading

8

We hope that this script has awakened your interest in data analysis. Of course, this is only a first introduction and there is still much to discover.

At this point, we would like to give a few references. The list is certainly not exhaustive and there are many other good books.

Generally, it should be mentioned that not every book is equally suitable for each student. We should look for literature and use books that fit our taste. In the following, the literature is sorted according to statistics and R books, whereby the classification is not completely precise.

With regard to statistics, the following books may be of interest:

Bortz, J.; Schuster, Ch. (2010), Statistik für Human- und Sozialwissenschaftler, 7. Auflage. Springer: Berlin Heidelberg.

Backhaus, K.; Erichson, B.; Plinke, W.; Weiber, R. (2011), Multivariate Analysemethoden, 13. Auflage. Springer: Berlin Heidelberg.

Eckey, H.F.; Kosfeld, R; Dreger, Ch. (2011), Ökonometrie: Grundlagen - Methoden - Beispiele, 4. Auflage. Gabler: Wiesbaden.

Greene, W.H: (2012), Econometric Analysis, 7th Edition. Pearson: Harlow, GB.

Hair, J.F.; Black, W.C.; Babin, B.J.; Aderson, R.E. (2013), Multivariate Data Analysis, 7th Edition. Pearson: Harlow, GB.

Gujarati, D. (2011), Econometrics by Example. Palgrave Macmillan: Basingstoke.

Kronthaler, F. (2016), Datenanalyse ist (k)eine Kunst. Berlin Heidelberg: Springer.

Salkind, N. (2011): Statistics for People Who (Think They) Hate Statistics. London: Sage Publications.

There are also many good books and scripts about R, e.g.:

Cotton, R. (2013): Learning R. Sebastopol CA: O'Reilly Media.

Crawley, M.J. (2013): The R Book. Chichester: Wiley & Sons.

Dalgaard, P. (2008): Introductory Statistics with R. New York: Springer-Verlag.

Faes, G. (2010): Einführung in R, Ein Kochbuch zur statistischen Datenanalyse. Norderstedt: BoD.

Field, A.; Miles, J.; Field, Z. (2012): Discovering Statistics Using R. London: Sage Publications.

© The Author(s), under exclusive license to Springer-Verlag GmbH, DE, part of Springer Nature 2021
F. Kronthaler and S. Zöllner, *Data Analysis with RStudio*,
https://doi.org/10.1007/978-3-662-62518-7_8

Wickham, H. (2009): ggplot2: Elegant Graphics for Data Analysis. New York: Springer-Verlag.

Teetor, P. (2011): R Cookbook: Proven Recipes for Data Analysis, Statistics, and Graphics. Sebastopol CA: O'Reilly Media.

Ugarte, M.D.; Militino, A.F.; Arnholt, A.T. (2016): Probability and Statistics with R. Boca Raton: CRC Press.

Wollschläger, D. (2017): Grundlagen der Datenanalyse mit R: Eine anwendungsorientierte Einführung. Berlin: Springer Spektrum.

Appendix

9

9.1 Appendix 1: Questionnaire

See Fig. 9.1.

9.2 Appendix 2: Dataset tourism.xlsx Including Legend

See Figs. 9.2 and 9.3.

Visitor survey ski resort

Dear visitor of our ski resort,
we of the mountain railway **Heide** are always trying to optimize our ski area for you and to adjust it to your desires and needs. In order to succeed, we carry out the following survey.

It takes about 5 minutes to answer the questionnaire. Of course we treat your data confidentially and analyse the data anonymously.

Your answers are very important for the development of our ski resort. Thank you very much for your support!

A. Questions regarding your current holiday

1. Accomodation
In which accommodation are you staying during your current ski holiday?

1.01 ☐ less than 2 star hotel   1.02 ☐ 2 star hotel   1.03 ☐ 3 star hotel

1.04 ☐ 4 star hotel   1.05 ☐ 5 star hotel

2. Duration
The total duration of your current ski holiday is: 2 ☐☐☐ day(s)

3. How satisfied are you with the following offers of our ski resort? Please mark the satisfaction on the line with a cross, 0 means very unsatisfied and 100 very satisfied.

	0 very unsatisfied	100 very satisfied
Diversity of ski slopes:	3a ├───────────────────	──────────────────┤
Waiting time at ski lifts:	3b ├───────────────────	──────────────────┤
Quality of restaurants:	3c ├───────────────────	──────────────────┤
Satiscation with the resort in total:	3d ├───────────────────	──────────────────┤

4. Price/performance ratio
How do you think about the price of the ski pass?

4.01 ☐ expensive   4.02 ☐ apporpriate   4.03 ☐ cheap

5. Expenses per day
How much money do you spend per day and guest, including accommodation?:   5 _ _ _ _ _ CHF

Fig. 9.1 Questionnaire tourism

6. Recommendation

Would you recommend our ski area to your friends and relatives?

6.01 ☐ yes   6.02 ☐ rather yes   6.03 ☐ rather no   6.04 ☐ no

7. Future ski holiday

Is it likely that you spend your skiing holiday in our ski resort again next year?

7.01 ☐ yes   7.02 ☐ no

B Demographics

8. Sex   8.00 ☐ male   8.01 ☐ female

9. Country of origin 9.01 ☐ CH   9.02 ☐ GER 9.03 ☐ AUT 9.04 ☐ other country

10. Age   I am...   ☐☐   years old
                                       10

11. Highest level of education

11.01 ☐ Secondary school   11.02 ☐ A-Level   11.03 ☐ Bachelor

11.04 ☐ Master

Thank you very much for your valuable assistance!

C. Will be filled in through the inverviewer.

Internal Information: Interviewer: _____ Location: _____ Date:_____

Fig. 9.1 (continued)

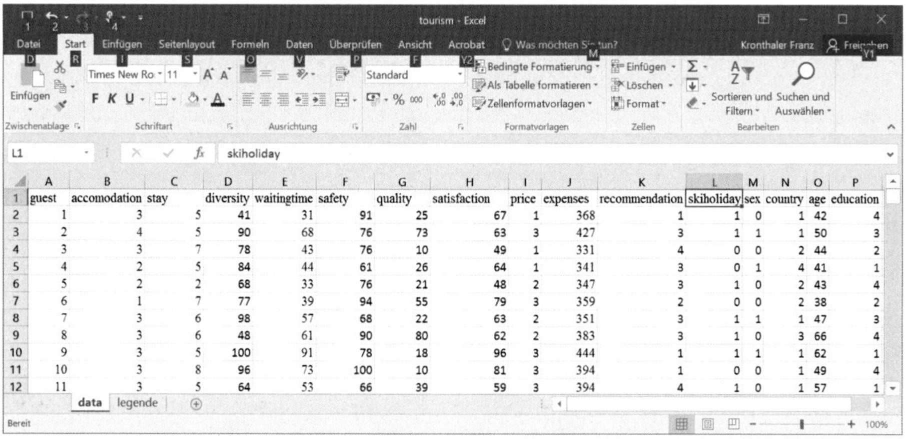

	A	B	C	D	E	F	G	H	I	J	K	L	M	N	O	P
1	guest	accomodation	stay	diversity	waitingtime	safety	quality	satisfaction	price	expenses	recommendation	skiholiday	sex	country	age	education
2	1	3	5	41	31	91	25	67	1	368	1	1	0	1	42	4
3	2	4	5	90	68	76	73	63	3	427	3	1	1	1	50	3
4	3	3	7	78	43	76	10	49	1	331	4	0	0	2	44	2
5	4	2	5	84	44	61	26	64	1	341	3	0	1	4	41	1
6	5	2	2	68	33	76	21	48	2	347	3	1	0	2	43	4
7	6	1	7	77	39	94	55	79	3	359	2	0	0	2	38	2
8	7	3	6	98	57	68	22	63	2	351	3	1	1	1	47	3
9	8	3	6	48	61	90	80	62	2	383	3	1	0	3	66	4
10	9	3	5	100	91	78	18	96	3	444	1	1	1	1	62	1
11	10	3	8	96	73	100	10	81	3	394	1	0	0	1	49	4
12	11	3	5	64	53	66	39	59	3	394	4	1	0	1	57	1

Fig. 9.2 Dataset tourism

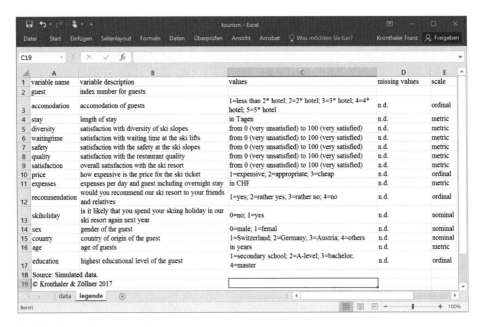

Fig. 9.3 Legend dataset tourism

9.3   Appendix 3: How to Deal with Missing Data

Whenever we collect data ourselves or use secondary data, it might happen that we have missing values. If, for example, a participant in our survey does not answer a question, this value is missing. How do we deal with it?

There are basically two ways how to deal with missing data. We can exclude the cases in which values are missing or replace the missing values with estimated values.

Missing values are often designated as numbers, e.g. 99, in statistics software, but in R they are encoded with NA. Hence, we have to convert the missing values, e.g. the 99, to NA when importing the data into RStudio. As always, there are several ways to do this. For example, we use the *naniar* package. After we installed the package the command is

```
> library(naniar)

> Datensatz <- Datensatz %>% replace_with_na(list(Variable=99))
```

Let us have a look at the first possibility. We want to exclude the cases in which values are missing. For demonstration purposes, we use the dataset *tourism_missing.xlsx*. Here we have deleted the first three entries for the variable *age*. We import the dataset into RStudio as usual and we try to calculate the mean of the variable *age* as usual. However, it does not work and R gives NA as the mean. Try it yourself.

```
> mean(tourism_missing$age)
```

Instead, we have to exclude the NA. The command is na.rm = TRUE which simply means to remove NAs. We learn that the mean age amongst the respondents is around 48 years.

```
> mean(tourism_missing$age, na.rm = TRUE)
```

Now we want to calculate the correlation between the satisfaction with the ski resort and the expenses in CHF per day. In our dataset *tourism_missing.xlsx* five entries for each variable are missing. Similar to the calculation of the mean, the calculation of the correlation is not possible due to the NA. Try it yourself. R gives NA as the correlation. We have to exclude the NAs here, too. However, the command *na.rm = TRUE* does not work in this case. Instead, we specify that we only want to use complete observations with *use = "complete.obs"*. We learn that the (Pearson) correlation is 0.4389338.

```
> cor(tourism_missing$satisfaction,tourism_missing$expenses, use =
"complete.obs")
```

We can also create a new dataset without the NA. This means we delete the whole answer from a respondent with a missing value. The command is na.omit(). We apply it to our data frame and call the new data frame *tourism_complete*. In our new data frame *tourism_complete*, the complete answers from respondents with missing values are deleted. In the data and object window at the top right, we see that the new data frame consists of only 138 observations instead of 150 observations in the original data frame.

```
> tourism_complete <- na.omit(tourism_missing)
> tourism_complete
```

The decision whether to delete the missing cases or replace the missing values must be answered individually. Let's assume that we delete all cases in a questionnaire, i.e. persons who have not entered their age. However, it is possible that age is related to gender, for example, that women do not like to give their age so much. If we delete the cases without age, we will have fewer women and therefore a biased sample.

Regarding the second possibility, to replace missing values with estimated values, we would like to refer the interested reader to the statistical literature.

9.4   Appendix 4: Solutions for the Tasks

Please note! Here the commands are shown to produce the solutions of the tasks, not the solutions itself. You can perform the commands to receive the solutions. However, you should only do so to check your own solutions, otherwise it will not help you to master the commands. Furthermore, the commands will only work if you use the correct path, in which you stored the data. Then a last comment with regard to the provided solutions. We did not test in every case the assumptions of the performed methods. However, you should do it and you may have then to adjust your commands accordingly.

Chapter 2

2.1

```
(5 + 2)*(10 − 3)
sqrt(64)
(64)^(1/3)
(512)^(1/3)
4^3
8^3
log(500)
log(10)
exp(1)
exp(5)
```

2.3

```
library(readxl)
dogs <- read_excel("D:/serie_r/tourismus/dogs.xlsx")
rm(dogs)
dogs <- read_excel("D:/serie_r/tourismus/dogs.xlsx")
rm(list = ls())
```

2.4

```
rm(list = ls())
library(readxl)
dogs <- read_excel("D:/serie_r/tourismus/dogs.xlsx")
head(dogs)
tail(dogs)
View(dogs)
str(dogs)
```

2.5

```
rm(list = ls())
library(readxl)
dogs <- read_excel("D:/serie_r/tourismus/dogs.xlsx")
dogs_red <- dogs[1:6,1:5]
```

2.6

```
rm(list = ls())
library(readxl)
dogs <- read_excel("D:/serie_r/tourismus/dogs.xlsx")
```

```
dogs_red <- dogs[1:6,1:5]
dogs_red[4,3]
View(dogs_red)
```

2.7

```
rm(list = ls())
library(readxl)
dogs <- read_excel("D:/serie_r/tourismus/dogs.xlsx")
humanage <- dogs$age*5
f_sex <- as.factor(dogs$sex)
f_size <- as.factor(dogs$size)
f_breed <- as.factor(dogs$breed)
dogs <- cbind(dogs, humanage, f_sex, f_size, f_breed)
```

2.8

```
# ———————————————————————————————————
# Title: First small script
# ———————————————————————————————————
# Author: …
# Date: …
# Preparation

rm(list = ls())
# do not forget to set your working directory with setwd()

# Read and check data

library(readxl)
dogs <- read_excel("D:/serie_r/tourismus/dogs.xlsx")
head(dogs)
tail(dogs)
View(dogs)
str(dogs)

# Prepare data

f_sex <- as.factor(dogs$sex)
dogs <- cbind(dogs, f_sex)

# Analysis

summary(dogs);
hist(dogs$age);
library(RcmdrMisc)
numSummary(dogs[,"age",   drop = FALSE],   statistics = c("mean",   "sd",   "IQR",
"quantiles"), quantiles = c(0,.25,.5,.75,1));
```

```
boxplot(dogs$age ~ dogs$f_sex);
numSummary(dogs[,"age",   drop = FALSE],   groups = dogs$f_sex,   statistics = c
("mean", "sd", "IQR", "quantiles"), quantiles = c(0,.25,.5,.75,1))
# ———————————————-
#End
```

Chapter 4

4.1

```
rm(list = ls())
library(readxl())
tourism <- read_excel("D:/serie_r/tourismus/tourism.xlsx")
table(tourism$accommodation)
table(tourism$education)
table(tourism$stay)
```

4.2

```
rm(list = ls())
library(readxl())
tourism <- read_excel("D:/serie_r/tourismus/tourism.xlsx")
median(tourism$age)
mean(tourism$age)
median(tourism$expenses)
mean(tourism$expenses)
median(tourism$stay)
mean(tourism$stay)
```

4.3

```
rm(list = ls())
library(readxl())
tourism <- read_excel("D:/serie_r/tourismus/tourism.xlsx")
sd(tourism$age)
var(tourism$age)
quantile(tourism$age)
sd(tourism$expenses)
var(tourism$expenses)
quantile(tourism$expenses)
sd(tourism$stay)
var(tourism$stay)
quantile(tourism$stay)
```

4.4

```
rm(list = ls())
library(readxl())
tourism <- read_excel("D:/serie_r/tourismus/tourism.xlsx")
library(RcmdrMisc)
numSummary(tourism[,c("diversity", "quality", "safety", "satisfaction", "waitingtime"),
drop = FALSE], statistics = c("mean", "sd", "var", "IQR", "quantiles"), quantiles =
c(0,.25,.5,.75,1))
```

4.5

```
rm(list = ls())
library(readxl())
tourism <- read_excel("D:/serie_r/tourismus/tourism.xlsx")
library(RcmdrMisc)
f_sex <-as.factor(tourism$sex)
tourism <- cbind(tourism, f_sex)
numSummary(tourism[,"waitingtime",   drop = FALSE],   groups = tourism$f_sex,
statistics = c("man", "sd", "IQR", "quantiles"), quantiles = c(0,.25,.5,.75,1))
```

4.6

```
rm(list = ls())
library(readxl())
tourism <- read_excel("D:/serie_r/tourismus/tourism.xlsx")
cor(tourism$age, tourism$expenses, method =  "pearson", use =  "complete")
```

4.7

```
rm(list = ls())
library(readxl())
tourism <- read_excel("D:/serie_r/tourismus/tourism.xlsx")
cor(tourism[,c("age","expenses","stay")], method = "pearson", use =  "complete")
```

4.8

```
rm(list = ls())
library(readxl())
tourism <- read_excel("D:/serie_r/tourismus/tourism.xlsx")
cor(tourism$accommodation,  tourism$education,  method =   "spearman",  use =
"complete")
```

4.9

```
rm(list = ls())
library(readxl())
tourism <- read_excel("D:/serie_r/tourismus/tourism.xlsx")
```

```
cor(tourism[,c("accommodation", "recommendation", "education")], method = "pear-
man", use = "complete")
```

4.10

```
rm(list = ls())
library(readxl())
tourism <- read_excel("D:/serie_r/tourismus/tourism.xlsx")
hist(tourism$expenses, main = "Expenses in categories", xlab = "Expenses", ylab =
"Frequency", col = "purple")
```

4.11

```
rm(list = ls())
library(readxl())
tourism <- read_excel("D:/serie_r/tourismus/tourism.xlsx")
pie(table(tourism$education), main = "Education of guests", labels = c("scondary
school", "A-Level", "bachelor", "master"))
```

4.12

```
rm(list = ls())
library(readxl())
tourism <- read_excel("D:/serie_r/tourismus/tourism.xlsx")
table(tourism$country, tourism$accommodation)
barplot(table(tourism$country, tourism$accommodation), main = "Accommodation
category by guests", xlab = "Accommodation category", ylab = "Frequency",
names.arg = c("less  than  2*",  "2*","3*","4*","5*"),  beside = TRUE,  col =
c("drkblue","red",  "green",  "yellow"),  legend = rownames(table(tourism$country,
tourism$accommodation)))
```

4.13

```
rm(list = ls())
library(readxl())
tourism <- read_excel("D:/serie_r/tourismus/tourism.xlsx")
boxplot(tourism$expenses ~ tourism$country, main = "Variation of the variable
expenses by country", xlab = "Country", ylab = "Expenses", names = c("Switzerland",
"Germany", "Austria", "others"))
```

4.14

```
rm(list = ls())
library(readxl())
tourism <- read_excel("D:/serie_r/tourismus/tourism.xlsx")
plot(tourism$age, tourism$expenses, main = "Relationship between age and expen-
ses", xlab = "Age", ylab = "Expenses in CHF per day")
```

4.15

```
rm(list = ls())
library(readxl())
tourism <- read_excel("D:/serie_r/tourismus/tourism.xlsx")
scatter.smooth(tourism$age, tourism$expenses, main =  "Relationship between age
and expenses", xlab =  "Age", ylab =  "Expenses in CHF per day")
```

Chapter 5

5.1

```
rm(list = ls())
library(readxl())
dogs <- read_excel("D:/serie_r/tourismus/dogs.xlsx")
hist(dogs$age)
library(car)
qqPlot(dogs$age)
library(e1071)
kurtosis(dogs$age)
skewness(dogs$age)
shapiro.test(dogs$age)
```

5.2

```
rm(list = ls())
library(readxl())
tourism <- read_excel("D:/serie_r/tourismus/tourism.xlsx")
hist(tourism$expenses)
library(car)
qqPlot(tourism$expenses)
library(e1071)
kurtosis(tourism$expenses)
skewness(tourism$expenses)
shapiro.test(tourism$expenses)
```

5.3

```
rm(list = ls())
library(readxl())
tourism <- read_excel("D:/serie_r/tourismus/tourism.xlsx")
hist(tourism$satisfaction)
library(car)
qqPlot(tourism$satisfaction)
library(e1071)
kurtosis(tourism$satisfaction)
```

```
skewness(tourism$satisfaction)
shapiro.test(tourism$satisfaction)
```

Chapter 6

6.1

```
rm(list = ls())
library(readxl())
tourism <- read_excel("D:/serie_r/tourismus/tourism.xlsx")
t.test(tourism$satisfaction, alternative =  "greater", mu = 60, conf.level = 0.95)
```

6.2

```
rm(list = ls())
library(readxl())
tourism <- read_excel("D:/serie_r/tourismus/tourism.xlsx")
f_sex <- as.factor(tourism$sex)
tourism <- cbind(tourism, f_sex)
t.test(expenses ~ f_sex,   data = tourism,   alternative = "less",   mu = 0,   paired =
FALSE, var.equal = TRUE, conf.level = 0.95)
```

6.3

```
rm(list = ls())
library(readxl())
tourism <- read_excel("D:/serie_r/tourismus/tourism.xlsx")
f_sex <- as.factor(tourism$sex)
tourism <- cbind(tourism, f_sex)
tapply(tourism$education, tourism$f_sex, median, na.rm = TRUE)
wilcox.test(education ~ f_sex, data = tourism, alternative = "two.sided")
```

6.4

```
rm(list = ls())
library(readxl())
schoolbreak <- read_excel("D:/serie_r/tourismus/schoolbreak.xlsx")
schoolbreak_red <- schoolbreak[1:8,]
diff <- schoolbreak_red$before-schoolbreak_red$after
schoolbreak_red <- cbind(schoolbreak_red, diff)
hist(schoolbreak_red$diff)
library(car)
qqPlot(schoolbreak_red$diff)
library(e1071)
kurtosis(schoolbreak_red$diff)
skewness(schoolbreak_red$diff)
shapiro.test(schoolbreak_red$diff)
```

```
t.test(schoolbreak_red$before, schoolbreak_red$after, alternative =  "greater", mu = 0,
paired = TRUE, conf.level = 0.95)
median(schoolbreak_red$before)
median(schoolbreak_red$after)
wilcox.test(schoolbreak_red$before,   schoolbreak_red$after,   alternative = "greater",
paired = TRUE)
```

6.5

```
rm(list = ls())
library(readxl())
tourism <- read_excel("D:/serie_r/tourismus/tourism.xlsx")
f_education <- as.factor(tourism$education)
tourism <- cbind(tourism, f_education)
hist(tourism$expenses)
library(car)
qqPlot(tourism$expenses)
shapiro.test(tourism$expenses)
Boxplot(expenses ~ f_education, data = tourism)
with(tourism, tapply(expenses, f_education, var, na.rm = TRUE))
leveneTest(expenses ~ f_education, data = tourism, center = "mean")
AnovaModel.1 <- aov(expenses ~ f_education, data = tourism)
summary(AnovaModel.1)
TukeyHSD(AnovaModel.1)
library(RcmdrMisc)
with(tourism, plotMeans(expenses, f_education, error.bars = "conf.int", level = 0.95))
```

6.6

```
rm(list = ls())
library(readxl())
tourism <- read_excel("D:/serie_r/tourismus/tourism.xlsx")
f_education <- as.factor(tourism$education)
f_sex <- as.factor(tourism$sex)
tourism <- cbind(tourism, f_education, f_sex)
hist(tourism$expenses)
library(car)
qqPlot(tourism$expenses)
shapiro.test(tourism$expenses)
Boxplot(expenses ~ f_education, data = tourism)
with(tourism, (tapply(expenses, list(f_education, f_sex), var, na.rm = TRUE)))
leveneTest(expenses ~ f_education*f_sex, data = tourism, center =  "mean")
AnovaModel.1 <- aov(expenses ~ f_education*f_sex, data = tourism)
summary(AnovaModel.1)
```

```
TukeyHSD(AnovaModel.1)
library(RcmdrMisc)
with(tourism, plotMeans(expenses, f_education, f_sex, error.bars =  "none"))
```

6.7

```
rm(list = ls())
library(readxl())
tourism <- read_excel("D:/serie_r/tourismus/tourism.xlsx")
plot(tourism$diversity, tourism$satisfaction)
plot(tourism$waitingtime, tourism$satisfaction)
plot(tourism$safety, tourism$satisfaction)
plot(tourism$quality, tourism$satisfaction)
with(tourism, cor.test(diversity, satisfaction, alternative =  "two.sided", method =
"pearson"))
with(tourism, cor.test(waitingtime, satisfaction, alternative = "two.sided", method =
"pearson"))
with(tourism,  cor.test(safety,  satisfaction,  alternative = "two.sided",  method =
"pearson"))
with(tourism,  cor.test(quality,  satisfaction,  alternative = "two.sided",  method =
"pearson"))
```

6.8

```
rm(list = ls())
library(readxl())
tourism <- read_excel("D:/serie_r/tourismus/tourism.xlsx")
with(tourism,  cor.test(education,  accommodation,  alternative =   "two.sided",
method =  "spearman", exact = FALSE))
```

6.9

```
rm(list = ls())
library(readxl())
tourism <- read_excel("D:/serie_r/tourismus/tourism.xlsx")
f_skiholiday <- as.factor(tourism$skiholiday)
f_accommodation <- as.factor(tourism$accommodation)
tourism <- cbind(tourism, f_skiholiday, f_accommodation)
table(tourism$f_skiholiday, tourism$f_accommodation)
chisq.test(tourism$f_skiholiday, tourism$f_accommodation)
```

6.10

```
rm(list = ls())
library(readxl())
tourism <- read_excel("D:/serie_r/tourismus/tourism.xlsx")
```

```
f_skiholiday <- as.factor(tourism$skiholiday)
f_country <- as.factor(tourism$country)
tourism <- cbind(tourism, f_skiholiday, f_country)
table(tourism$f_skiholiday, tourism$f_country)
chisq.test(tourism$f_skiholiday, tourism$f_country)
```

Chapter 7

7.1

```
rm(list = ls())
library(readxl())
tourism <- read_excel("D:/serie_r/tourismus/tourism.xlsx")
plot(tourism$expenses, tourism$age)
RegModel.1 <- lm(expenses ~ age, data = tourism)
summary(RegModel.1)
tourism <- cbind(tourism, RegModel.1$residuals)
plot(tourism$age, RegModel.1$residuals)
library(ggplot2)
library(ggfortify)
autoplot(RegModel.1)
library(olsrr)
ols_test_breusch_pagan(RegModel.1)
```

7.2

```
rm(list = ls())
library(readxl())
tourism <- read_excel("D:/serie_r/tourismus/tourism.xlsx")
pairs( ~ expenses + age + stay + satisfaction,   data = tourism,   main =    "Simple
Scatterplot Matrix")
RegModel.2 <- lm(expenses ~ age + stay + satisfaction, data = tourism)
summary(RegModel.2)
tourism <- cbind(tourism, RegModel.2$residuals)
pairs( ~ RegModel.2$residuals + age + stay + satisfaction,   data = tourism,   main =
"Simple Scatterplot Matrix")
library(ggplot2)
library(ggfortify)
autoplot(RegModel.2)
library(olsrr)
ols_test_breusch_pagan(RegModel.2)
cor(tourism[,c("age", "stay", "satisfaction")])
library(RcmdrMisc)
vif(RegModel.2)
```

7.3

```
rm(list = ls())
library(readxl())
tourism <- read_excel("D:/serie_r/tourismus/tourism.xlsx")
library(fastDummies)
tourism <- dummy_cols(tourism, select_columns = c("accommodation"))
colnames(tourism)[colnames(tourism)=="accommodation_1"] <- "smaller"
colnames(tourism)[colnames(tourism)=="accommodation_2"] <- "two_star"
colnames(tourism)[colnames(tourism)=="accommodation_3"] <- "three_star"
colnames(tourism)[colnames(tourism)=="accommodation_4"] <- "four_star"
colnames(tourism)[colnames(tourism)=="accommodation_5"] <- "five_star"
pairs( ~ expenses + age + stay + satisfaction, data = tourism, main =  "Simple Scat-
terplot Matrix")
RegModel.3 <-lm(expenses  ~  age + stay + satisfaction + two_star + three_star +
four_star + five_star, data = tourism)
summary(RegModel.3)
tourism <- cbind(tourism, RegModel.3$residuals)
pairs( ~ RegModel.3$residuals + age + stay + satisfaction,   data = tourism,   main =
"Simple Scatterplot Matrix")
library(ggplot2)
library(ggfortify)
autoplot(RegModel.3)
library(olsrr)
ols_test_breusch_pagan(RegModel.3)
cor(tourism[,c("age", "stay", "satisfaction")])
library(RcmdrMisc)
vif(RegModel.3)
```

7.4

```
rm(list = ls())
library(readxl())
tourism <- read_excel("D:/serie_r/tourismus/tourism.xlsx")
library(fastDummies)
tourism <- dummy_cols(tourism, select_columns = c("accommodation"))
colnames(tourism)[colnames(tourism) == "accommodation_1"] <- "smaller"
colnames(tourism)[colnames(tourism) == "accommodation_2"] <- "two_star"
colnames(tourism)[colnames(tourism) == "accommodation_3"] <- "three_star"
colnames(tourism)[colnames(tourism) == "accommodation_4"] <- "four_star"
colnames(tourism)[colnames(tourism) == "accommodation_5"] <- "five_star"
pairs( ~ expenses + age + stay + satisfaction,   data = tourism,   main =    "Simple
Scatterplot Matrix")
```

```
RegModel.4 <-  lm(expenses ~ age + sex + stay + satisfaction + two_star + three_
star + four_star + five_star, data = tourism)
summary(RegModel.4)
tourism <- cbind(tourism, RegModel.4$residuals)
pairs( ~ RegModel.4$residuals + age + stay + satisfaction,  data = tourism,  main =
"Simple Scatterplot Matrix")
library(ggplot2)
library(ggfortify)
autoplot(RegModel.4)
library(olsrr)
ols_test_breusch_pagan(RegModel.4)
cor(tourism[,c("age", "stay", "satisfaction")])
library(RcmdrMisc)
vif(RegModel.4)
```

Printed in the United States
By Bookmasters